P9-CJN-627

BIOMASS CONVERSION TECHNOLOGY

PRINCIPLES and PRACTICE

Pergamon Titles of Related Interest

Braunstein BIOMASS ENERGY SYSTEMS AND THE ENVIRONMENT

Fiksel BIOTECHNOLOGY RISK ASSESSMENT

Goodman BIOMASS ENERGY PROJECTS, PLANNING AND MANAGEMENT

ICE EFFLUENT TREATMENT IN THE PROCESS INDUSTRIES

Johnson MODELLING AND CONTROL OF BIOTECHNOLOGICAL PROCESSES

Moo-Young COMPREHENSIVE BIOTECHNOLOGY, 4 Volumes

Moo-Young WASTE TREATMENT AND UTILIZATION

OTA COMMERCIAL BIOTECHNOLOGY

Perpich BIOTECHNOLOGY IN SOCIETY

Stanbury PRINCIPLES OF FERMENTATION TECHNOLOGY

Related Journals

(Sample copies available on request)

BIOTECHNOLOGY ADVANCES

THE CHEMICAL ENGINEER

CHEMICAL ENGINEERING SCIENCE

ENERGY

ENERGY CONVERSION AND MANAGEMENT

ENVIRONMENT INTERNATIONAL

NUCLEAR AND CHEMICAL WASTE MANAGEMENT

OUTLOOK ON AGRICULTURE

WATER RESEARCH

WATER SCIENCE AND TECHNOLOGY

BIOMASS CONVERSION TECHNOLOGY

PRINCIPLES and PRACTICE

Editor

M. Moo-Young

Department of Chemical Engineering
University of Waterloo, Canada

Associate Editors

J. Lamptey

Pioneer Hi-Bred, U.S.A.

B. Glick

University of Waterloo, Canada

H. Bungay

Rensselaer Polytechnic Institute, U.S.A.

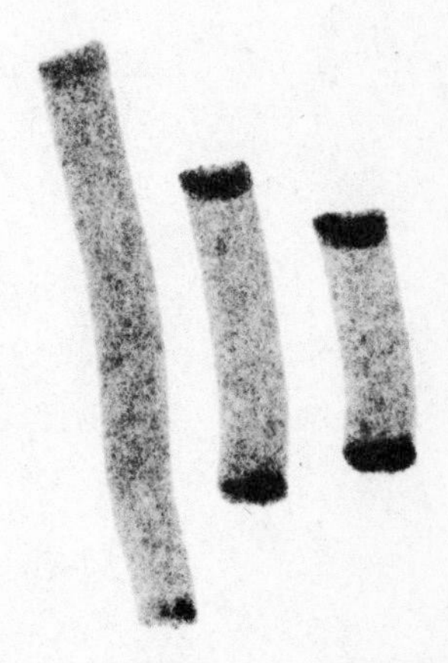

PERGAMON PRESS

New York Oxford Beijing Frankfurt
São Paulo Sydney Tokyo Toronto

Pergamon Press Offices:

U.S.A.	Pergamon Press, Maxwell House, Fairview Park, Elmsford, New York 10523, U.S.A.
U.K.	Pergamon Press, Headington Hill Hall, Oxford OX3 0BW, England
PEOPLE'S REPUBLIC OF CHINA	Pergamon Press, Room 4037, Qianmen Hotel, Beijing, People's Republic of China
FEDERAL REPUBLIC OF GERMANY	Pergamon Press, Hammerweg 6, D-6242 Kronberg, Federal Republic of Germany
BRAZIL	Pergamon Editora, Rua Eça de Queiros, 346, CEP 04011, Paraiso, São Paulo, Brazil
AUSTRALIA	Pergamon Press (Aust.) Pty., P.O. Box 544, Potts Point, NSW 2011, Australia
JAPAN	Pergamon Press, 8th Floor, Matsuoka Central Building, 1-7-1 Nishishinjuku, Shinjuku-ku, Tokyo 160, Japan
CANADA	Pergamon Press Canada Ltd, Suite 271, 253 College St, Toronto, Ontario, Canada M5 T1RS

First printing 1987

Library of Congress Cataloging in Publication Data
Biomass conversion technology.
Includes index.
1. Biomass chemicals. 2. Biotechnology.
I. Moo-Young, Murray.
TP248.B55B55 1986 662'.8 86-25370
ISBN 0-08-033174-2

In order to make this volume available as economically and as rapidly as possible, the authors' typescripts have been reproduced in their original forms. This method unfortunately has its typographical limitations but it is hoped that they in no way distract the reader.

Printed in Great Britain by A. Wheaton & Co. Ltd., Exeter

Table of Contents

Section 4: PRODUCTION AND ACTION OF CELLULASES

Section 5: OTHER BIOMASS CONVERSION TECHNOLOGIES

PREFACE

Biomass in the form of forestry, agricultural and agro-industrial materials, represents the largest renewable resource in the world. Unlike petroleum, biomass is well-distributed globally. However, this biomass resource is presently under-utilized and, in fact, is often regarded as surplus or waste material.

The basic chemistry of these diverse biomass materials is similar: polymers of five- and six-carbon sugar polymers (hemicellulose and cellulose) and aromatic ring polymers (lignin). An array of industrially important products can be produced from these polymers by various conversion technologies. New and improved techniques, many capitalizing on integrated physical, chemical and biological process strategies, are being developed to produce a wide range of useful products. Efforts are also being directed at improved methods of cultivating biomass materials on the one hand, and of removing them as potential environmental pollutants on the other.

In this monograph, all these aspects of biomass production and utilization are considered, including fundamental principles as well as practical applications. Scientists, engineers and others who are interested in learning or reviewing some of the basics and current developments in biomass conversion technologies are addressed. The manuscripts are organized under the following five sections:

1. Biomass Pretreatment
2. Production of Fuels and Solvents
3. Production of SCP
4. Production and Action of Cellulases
5. Other Biomass Conversion Technologies

In the preparation of this monograph, invaluable assistance in the reviewing of manuscripts was obtained from my associate editors: Jonathan Lamptey, Bernie Glick and Henry Bungay. Proof-reading and production co-ordination was provided by Arlene Lamptey. Finally, I wish to acknowledge the financial support of the Natural Sciences and Engineering Research Council of Canada and UNESCO in this publication.

Waterloo, Ontario, Canada — Murray Moo-Young

June 1986

Section 1

Biomass Pretreatment

OLIGOSACCHARIDES PRODUCED BY TREATING WOOD CHIPS WITH GASEOUS HYDROFLUORIC ACID

Claudio A. Chuaqui and John Merritt

Medical Biophysics Branch
Atomic Energy of Canada Limited Research Company
Whiteshell Nuclear Research Establishment
Pinawa, Manitoba Canada R0E 1L0

ABSTRACT

HPLC analysis of aqueous extracts of poplar wood chips, pretreated with gaseous hydrofluoric acid, shows the presence of oligosaccharides of low molecular weight. The oligosaccharides, which appear to have less than 10 units per chain, are readily extractable from the HF-treated chips and account for more than 85% of the theoretically available sugars. The monosaccharides are formed by acid hydrolysis under mild conditions in almost quantitative yields.

The results suggest that fermentable sugars from wood chips, pretreated with HF(g), may be obtained by subjecting the chips to a relatively mild acid hydrolysis. Alternatively, equally satisfactory results are produced after exhaustive extraction, followed by hydrolysis of the resulting homogeneous solution.

INTRODUCTION

At the end of 1981, the Ethanol-from-Cellulose program was implemented and assigned to Canertech Inc. with headquarters in Winnipeg. The main purpose of the program is to establish an optimum procedure to produce ethanol from lignocellulosics. Thus far, it has used poplar wood chips as the feedstock to produce fermentable sugars by the action of anhydrous hydrofluoric acid (HF) on the lignocellulosic material, followed by a secondary treatment. Although the action of HF on cellulose has been known for decades and early applications of it occurred in Germany in 1938, a complete process was never implemented to the stage of a pilot plant (1).

Presently, the use of HF in treating lignocellulosics has been adopted by several groups, including the Ethanol-from-Cellulose program (2-6). So far, the results obtained by this program are encouraging and strongly suggest that anhydrous HF may be used as the primary reagent to obtain fermentable sugars from wood chips at pilot plant scale.

EXPERIMENTAL

Poplar wood chips were treated with gaseous hydrofluoric acid as described previously (6).

The hydrolysis of HF-treated wood chips and aqueous extract solutions were carried for times that varied between 30 to 60 minutes at temperatures between 110°C and 140°C, and with sulfuric acid concentrations between 0.5 and 1.0%.

The extraction of oligosaccharides was performed by treating HF-treated wood chips with hot water. It was found that a temperature of about 80°C was sufficient to effect the extractions.

Sugars were analysed by HPLC techniques (8), using a Beckman instrument model 334 equipped with a Waters differential refractometer detector model R401, a Biorad HPX-87P cation exchange column (kept at 85°C) and Aminex guard columns. The solvent used was water at a flow rate of 0.6 mL/min.

RESULTS AND DISCUSSION

It has been demonstrated that gaseous hydrofluoric acid reacts with cellulose (I) to produce a glucopyranosyl fluoride derivative (III) via an oxonium ion intermediate (II). Removal of HF induces polymerization to yield oligosaccharides (IV) with molecular weights that depend on the experimental conditions employed (7). This process is illustrated in Figure 1, where it is shown that oligosaccharides are also produced from glucose (V), demonstrating the reversible nature of the reaction. The same general reaction mechanism applies also to other monosaccharides, including pentoses, which form oligomeric compounds when treated with HF(g).

Thus, when lignocellulosics such as wood chips are treated with HF(g), oligosaccharides are produced. However, the complexity of the material treated makes possible several undesired side reactions and the optimal reaction conditions changes accordingly. In addition, both cellulose (mainly hexosans) and hemicellulose (mainly pentosans) serve as substrates that yield oligosaccharides of varied structures.

Analysis of water extracts of poplar wood chips, pretreated with gaseous hydrofluoric acid as previously described (6) showed that the size of the oligosaccharides formed is quite small. The extracts were obtained by filtration, after treating the HF-treated wood chips with hot water (ca 80°C). These solutions were analysed by HPLC, using an HPX-87P Biorad column and as described in the Experimental. The chromatograms show a sharp single peak at a position that would correspond to a chain length of about nine thousand monosaccharide units, as seen in the upper HPLC chromatogram of Figure 2. These and other results, such as solubility characteristics, strongly suggest that the oligosaccharides produced contain less than ten sugar units. However, it is not possible to rule out the presence of oligosaccharides of higher molecular weight, which may not be adequately separated by the HPLC column used in the present work.

The structural characteristics of these oligosaccharides are complicated by two additional factors. The first one refers to their mixed monosaccharide nature, since there would be both hexoses (mainly glucose) and pentoses (mainly xylose) in their chains. The second factor refers to the mixed nature of linkages between monosaccharide units in each chain: it is expected that the beta-1-4 bond type, originally present in cellulose, as well as the readily hydrolysable alpha-1-6 bond would coexist in these oligomers.

Exhaustive extractions of HF-treated wood chips demonstrate that oligosaccharides are readily extractable. Calculations show that more than 85% of the theoretically available saccharides from the wood chips becomes water soluble as short-chain oligosaccharides. Moreover, these oligomers are readily hydrolysed, under mild acidic conditions, to yield monosaccharides in almost quantitative yields. This is illustrated in Figure 2, where the HPLC chromatogram at the bottom belongs to the extract solutions after undergoing acid hydrolysis.

CONCLUSIONS

The results obtained to date indicate that fermentable sugars from poplar wood chips, pretreated with gaseous hydrofluoric acid, may be obtained by different means. Thus, high yields of monosaccharides may be obtained by subjecting the chips to a relatively mild acid hydrolysis. Alternatively, equally satisfactory results are produced after exhaustive aqueous extraction, followed by hydrolysis of the resulting homogeneous clear solution.

REFERENCES

1. H. Luers (1938). Holz Roh-und Werkstoff, 1, 342-344

2. Sharkov, V.I., A.K. Bolotova and T.A. Boiko (1972). Kompleksnaya Pererabotka Rastitel'nogo Syr'ya, 39-49.

3. Selke, S.M., M.C. Hawley, H. Hardt, D.T.A. Lamport, G. Smith and J. Smith (1982). Ind. Eng. Chem. Prod. Res. Dev., 21, 11-16.

4. Franz, R., R. Erckel, T. Riehm, R. Woernle and H.M. Deger (1982). 9th Cellulose Conference, Syracuse, NY, USA.

5. Defaye, J., A. Gadelle, J. Papadopoulos and C. Pedersen (1982). 9th Cellulose Conference, Syracuse, NY, USA.

6. Ostrovski, C.M. (1984). Biomass Conversion Technology Symposium, Waterloo, Canada

7. Defaye, J., A. Gadells (1982). Carbohydrate Res., 110 217-227.

8. Wentz, F.E., A.D. Marcy and M.J. Gray (1982). Chromat. Sci., 20, 349-352.

ACKNOWLEDGEMENTS

This research is part of work performed under contract from the Ethanol-from-Cellulose Program, EM&R, Winnipeg, MB. The authors are grateful to Mr. J. Aitken and Mr. C. Ostrovski of the above program for useful discussions.

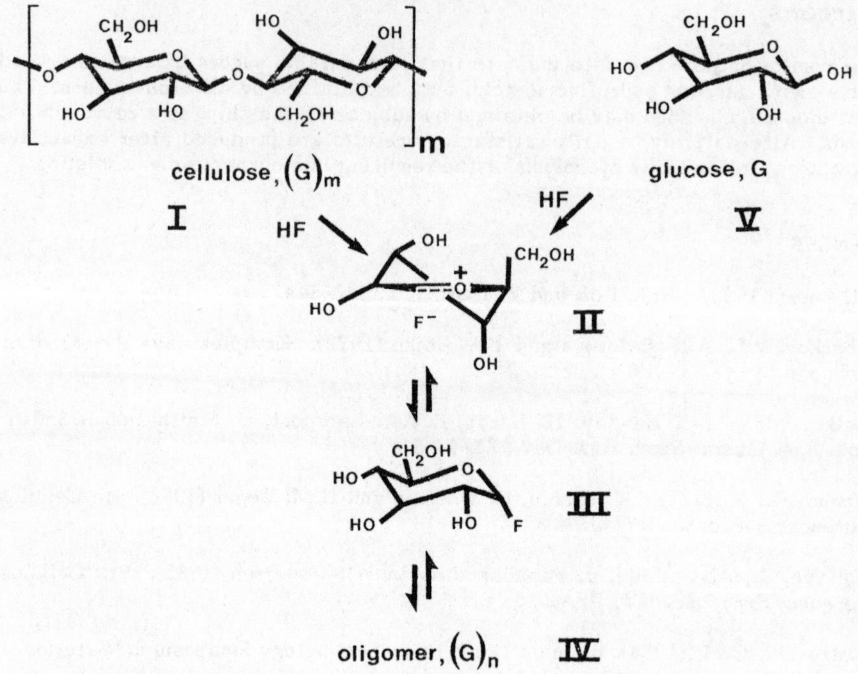

cellulose, $(G)_m$ I

glucose, G V

HF

II

III

oligomer, $(G)_n$ IV

Figure 1: Reaction of HF with cellulose (I) or glucose (V) to yield oligosaccharides of low molecular weight (IV) via intermediates II and III.

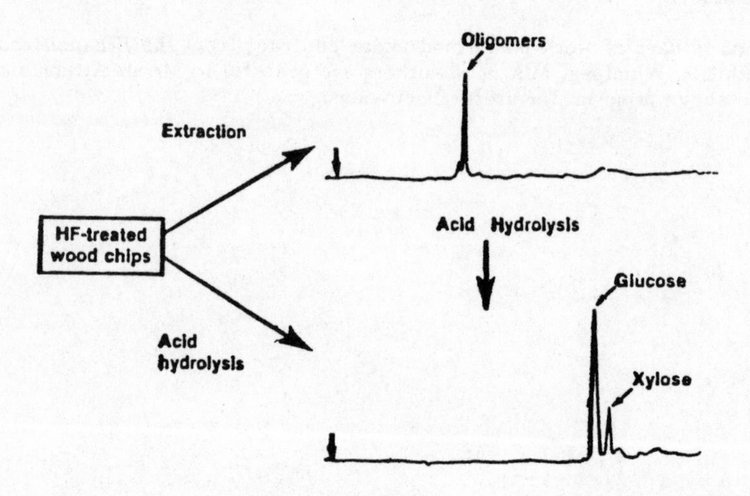

Figure 2: HPLC chromatograms of aqueous extract of HF-treated wood chips (top) and hydrolysate of HF-treated chips (bottom). The same chromatogram is also obtained after the aqueous extract is subjected to mild hydrolysis.

BIOTECHNOLOGY IN CROP PRODUCTION

A.I. de la Roche

Director, Chemistry & Biology Research Institute
Research Branch
Agriculture Canada
Ottawa, Ontario
K1A 0C6 Canada

This manuscript presents an overview of Agriculture Canada's biotechnology activities relating to crop improvement. While virtually all our efforts are directed towards cultivar development of traditional food and fibre crops, namely cereals, legumes and forages, the principles and approaches being used are equally applicable to energy crops like Jerusalem artichokes, Kochia and fodder beets, all of which are potential sources of biomass for fuel production. The strategy of any crop development program is to generate useful genetic variability, screen for the optimum combination of desirable traits, and then fix or stabilize this combination through several generations of inbreeding. While biotechnology is not expected to replace existing technology, it will certainly complement it. It will do this by making available new ways of generating genetic variation, by facilitating gene transfer and by providing new and more efficient systems for selection, thereby reducing substantially the time required to produce new varieties.

Applications of biotechnology in crop improvement involve the techniques of in vitro organ, tissue and cell culture, protoplast fusion and regeneration into hybrid plants as well as recombinant DNA methodology (Table 1). An integral part of each group of techniques listed is the ability to regenerate plants from cultured cells or tissues. It is important to note that many species, including most of the cereals and grain legumes, are generally recalcitrant to standard tissue culture protocols and thus are not at the present time amenable to cellular and molecular manipulations.

Some of the direct uses of in vitro culture includes asexual propagation of horticultural crops, elimination of viral diseases via meristem culture, preservation of valuable germ plasm generated asexually, generation of somaclonal variants and in vitro selection for economically important traits. Several of these techniques are already under commercial exploitation.

The production of haploid plants through another culture provides an efficient method for obtaining homozygous lines. This technique involves the culturing of the male reproductive structures found in flowers. Under appropriate conditions, developing pollen grains within the anthers can be induced to undergo plant regeneration. Plants obtained in this manner have been termed "haploids" because they possess only a single, rather than the normal double copy of the chromasomes, or units of DNA which form the genetic blueprint of the plant. Haploids can be utilized in crop improvement in a number of ways. These include the rapid breeding of homozygous diploid crops, the rapid development of pure lines for hybrid seed production and the section of novel genetic traits or mutations in cell culture. Desirable mutants would include those that are resistant to disease or tolerant to low temperature, drought, saline conditions, pollutants and herbicides.

Table 1 Biotechnology Objectives (Goals) in Relation to Crop Improvement

Objective	Technique(s)
Plant propagation	- meristem culture
Pathogen elimination/detection	- meristem culture
	- cell fusion (monoclonals)
Germplasm preservation	- protoplast, cell, meristem culture
Homozygous lines (haploids)	- anther, pollen, ovule, embryo culture
Genetic variants (mutants)	- protoplast, cell culture
	- mutagenesis
Interspecific hybrids	- embryo culture
	- cell fusion
Genetic transformation	- cell fusion
	- microinjection
	- rDNA methodology

Production of interspecific and intergeneric sexual hybrids in cereals is now possible through embryo culture techniques. At the Central Experimental Farm in Ottawa we have a well-established program in this area. It has led to the production and cytological characterization of a range of hybrids, including wheat x barley, barley x rye, and wheat or barley x several wild species. At our Ste. Foy Research Station, wheat x Agropyron hybrids have been obtained via embryo culture, an important step in our long term objective of transferring specific genes for viral resistance into cereals.

Protoplast fusion is another technique which can be employed in plant improvement. By fusing cells from different species it is possible to produce hybrid cells that can in turn be used to regenerate hybrid plants. Thus, the barriers which normally prevent cross fertilization between remotely related plant species can be bypassed and novel hybrids can be developed. We have generated a large number of tobacco interspecific hybrids which are being used at our tobacco breeding station in Delhi, Ontario. One particular hybrid cell fusion has resulted in a variant that is unusually high in nicotine, low in tar, and exhibits complete immunity to blue mold disease. This hybrid which is fertile has been backcrossed to common tobacco for several generations and a new variety is expected to be released with the next year.

Recently, we obtained seed from a somatic cell fusion of eggplant (Solanum melongena) which is a distant relative of S. sisymbriofolium. The latter carries resistance to root-knot nematode and Verticillium wilt. Derived cell lines are currently being screened for disease resistance in collaboration with colleagues at the U.S.D.A. A number of new somatic hybridization projects have been initiated in Agriculture Canada laboratories. One example is the cell fusion experiments between alfalfa and bird's food trefoil with the aim of transferring bloat resistance genes into alfalfa.

Plant genetic engineering, or recombinant DNA technology is the newest and most exciting area of plant biotechnology. Back in the late 1970's, it was shown that the T-region of the Ti plasmid from the soil bacterium Agrobacterium tumefaciens is inserted into the genome of infected plants. The inserted piece of DNA encodes for both opines production and tumor induction. This bacteria has been shown to naturally infect a broad spectrum of dicotyledonous plants. Transformed cells produce opines, undergo proliferation and possess the property of being autotrophic for growth hormones. In other words, they will continue to divide in vitro without an exogenous source of growth regulators. Recently, Monsanto scientists have used the Ti plasmid as a vector to transfer bacterial antibiotic resistance genes into plant cells of petunia. Mature plants have been regenerated and shown to be stably transformed. Genes for herbicide resistance and the small protein subunit of ribulase bisphosphate carboxylase have also been tranferred to plant cells using the Ti vector.

Investigators at Agriculture Canada in Ottawa have recently initiated a research project involving the genetic transformation of Brassica spp. by the Ti plasmid and are presently characterizing regenerants derived from transformed (crown gall) tissue. Another project established by the Ottawa group and involving collaboration with a group from Carleton University involves the development of microinjection techniques for introducing foreign DNA directly into plant cell nuclei. Within the past few months, this technique has been used to introduce Ti plasmid into alfalfa protoplasts. Evidence for transformation of microinjected cells (based on the detection of opine synthesis), has recently been obtained in a number of protoplast-derived colonies.

Gene transfer programs involving recombinant DNA techniques require major, long term commitments involving teams of skilled personnel. A major problem in Canada is the lack of scientists skilled in plant molecular genetics. Another general constraint in this field relates to our limited knowledge of gene expression and regulation in higher plants. At present, investigators have access to only a very limited number of plant genes for potential use in genetic engineering. Furthermore the technology, as it presently exists, has been established to transfer single genes. Unfortunately, many agronomic traits (i.e., yield, tolerance to environmental stress, etc.) are polygenically inherited. However, over the long term, plant breeding will certainly benefit immensely from the development and application of molecular techniques. Even now, molecular approaches to genome characterization, independent of genetic engineering, have been found to be powerful tools.

In summary, biotechnology could have a major impact on agriculture and food production, particularly in plant and animal strain improvement. To succeed the technology will have to be judiciously integrated into conventional technologies and approaches. We have already witnessed some impressive achievements with this new technology in terms of crop improvement, which today is deriving economic benefits for Canada.

KINETIC STUDIES OF WHEAT STRAW HYDROLYSIS USING SULPHURIC ACID

S. Ranganathan, N.N. Bakhshi, and D.G. Macdonald

Department of Chemical Engineering
University of Saskatchewan
Saskatoon, Saskatchewan, Canada, S7N 0W0

ABSTRACT

The kinetics of cellulose and hemicellulose hydrolysis of wheat straw were studied using both isothermal and non-isothermal techniques in a batch reactor. Reactions were carried out between 100 and 210°C and product sugars were analyzed using a Bio-Rad HPX-87P liquid chromatographic column. A simple first order series reaction model was used for both cellulose and hemicellulose hydrolysis reactions and kinetic parameters were obtained for the Arrhenius rate equations for three different sulphuric acid concentrations (0.5, 1.0 and 1.5%). Activation energies remained constant with acid concentration but the pre-exponential factors showed an increase with acid concentration. To minimize the amount of experimental data required and to achieve a unique solution to the kinetic parameters, the technique of combining isothermal and non-isothermal reaction data was studied.

INTRODUCTION

Wheat straw, containing 50% cellulose, 24% hemicellulose and 9.8% lignin, was treated with dilute sulphuric acid in a batch reactor at temperatures ranging from 100 to 210°C. Both isothermal and non-isothermal methods were used. The kinetic model used to analyze the data was a simple series reaction in which the cellulose or hemicellulose was hydrolyzed to monomeric sugars which in turn would degrade to degradation products. Only the initial cellulose/hemicellulose and the final glucose/xylose concentrations needed to be measured. The Arrhenius equation was used to describe the rate constants for each reaction (assumed first order) and the activation energies and pre-exponential factors for each reaction were determined by computer analysis from the experimental data.

KINETIC MODEL

The kinetics of acid hydrolysis of cellulose was first modelled by Saeman[1] who proposed a series reaction sequence as follows:

$$\text{Cellulose} \rightarrow \text{Glucose} \rightarrow \text{Degradation Products} \qquad (1)$$

The kinetics of hemicellulose hydrolysis was modelled by Mehlberg and Tsao[2] according to the following reaction sequence:

$$\text{Hemicellulose} \rightarrow \begin{cases} \text{Xylan I} \\ \text{Xylan II} \end{cases} \rightarrow \text{Oligomers} \rightarrow \text{Xylose} \rightarrow \begin{matrix} \text{Degradation} \\ \text{Products} \end{matrix} \quad (2)$$

Since the concentrations of Xylan I and II and oligomers are very difficult to determine and since the reaction to produce Xylan I and II is very rapid, it was decided to simplify the model to one that is similar to that for cellulose; that is,

$$\text{Hemicellulose} \rightarrow \text{Xylose} \rightarrow \text{Degradation Products} \quad (3)$$

The simple series reactions shown by equations 1 and 3 were assumed to be first order and their rate constants, k, can be represented by the Arrhenius form as follows:

$$k = A \cdot \exp(-E/RT) \quad (4)$$

where E is an activation energy which is independent of acid concentration while A is the pre-exponential factor which is a function of the acid concentration, usually reported as:

$$A = A_o \cdot C^m \quad (5)$$

EXPERIMENTAL APPARATUS AND PROCEDURE

Equipment

A 1-L Parr batch reactor (Parr Instrument Co., Moline, Illinois, 61265) with cooling coils, stirrer and heating jacket was used for the experiments. A thermocouple probe inside the reactor was connected to a digital temperature indicator and chart recorder. A stainless steel cup was welded to the shaft of the stirrer at a position above the liquid level. A high pressure liquid chromatograph (HPLC) with Bio-Rad HPX-87P chromatographic column was used for the sugar analysis. Column and eluent (deionized and degassed water) temperatures were maintained at 85°C. An Eldex B-100-S pump was used to pump the eluent and the sample through the HPLC column. A microguard supplied by Bio-Rad laboratories was used prior to the column in order to protect the column from damage by solid impurities in the sample. Sugars separated by the column were detected by a R-401 differential refractometer supplied by Waters Associates Inc. while a Hewlett Packard 3390A integrator was used for the integration of the signals obtained from the refractometer.

Non-Isothermal Method

Wheat straw, grown near Saskatoon, was pulverized in a Wiley pulverizer. Pulverized straw of -30 to +80 mesh size was dried in an oven overnight at 80-90°C and 15 g of it was placed in the reactor along with 300 mL of dilute sulphuric acid. The reactor was tightly closed and placed in the preheated jacket. The stirrer speed was adjusted to 900 rpm and the thermocouple probe was connected to the digital temperature indicator and chart recorder. The temperature-time history was recorded manually using the digital temperature indicator. The temperatures, which could not be recorded manually, were read later from the chart. The reaction was stopped by passing cooling water through the cooling coils in the reactor. The reaction mixture could be cooled to 100°C within one minute. The product mixture was filtered through Whatman (grade 202) filter paper without washing the filter cake, and the filtrate was analyzed for glucose and xylose using the HPLC column. Three different acid concentrations (0.5, 1.0, and 1.5%) were used for the experiments and five to six experiments were carried out for each concentration. In each case the reactor was heated from room temperature to a final temperature between 120 and 210°C. The heating rate of the reactants depended upon the degree of preheating of the heating jacket; thus, the temperature-time profiles were different for each of the experimental runs and this was taken into account in the computer analysis of the results.

Isothermal Method

In this method, 25 mL of sulphuric acid solution was placed in the cup attached to the stirrer. Straw and 275 mL of water were then added to the reactor. The reactor was sealed and heated to the desired temperature with the stirring speed at 200 rpm. Once the temperature was isothermal, the stirring speed was increased to 900 rpm to throw the acid out of the cup and into the straw slurry. The rest of the procedure was the same as the non-isothermal method.

Sugar Analysis

The liquid samples after acid hydrolysis had a very low pH and had to be adjusted to a pH between 5 and 7 before injection into the HPLC column. A 25 mL sample was placed in a beaker and small quantities of calcium hydroxide were added at a time until the pH reached 5. The sample was allowed to stand for 15 minutes so that any precipitate would settle. The supernatant liquid was transferred to another beaker and 0.6 to 0.7 g of deionizing resin (Amberlite monobed (MB-1)) was added to it. The sample was shaken for three minutes on a rotary shaker and filtered through 0.45 μm filter paper using a syringe type filtration unit supplied by Millipore Corporation. The sample was then injected into the HPLC column using a 50 μL loop. Eluent flow was kept at 0.4 mL/min.

Chemical Analysis of Wheat Straw

The wheat straw used in this study was analyzed for cellulose, hemicellulose and lignin according to the methods described by Goering and Van Soest[3] and the American Association of Cereal Chemists (AACC) established procedures.[4] The methods are based on the use of neutral and acid detergents for the removal of soluble carbohydrates, proteins and tannins. The neutral detergent fiber (NDF) method provides a measure of the total cell-wall material (cellulose, hemicellulose and lignin). The acid detergent fiber (ADF) method determines the cellulose plus lignin content. Consequently, the difference between NDF and ADF provides a determination of hemicellulose. Lignin may be found separately in the ADF fraction by removal of lignin by permanganate oxidation. Details of the method used have been given by Ranganathan.[5]

Computer Analysis of Data

The experimental data was used to calculate the pre-exponential factors and activation energies for each rate constant. A mathematical expression was obtained for the temperature-time history of each run and the rate equations for either the cellulose or hemicellulose reaction series were integrated separately for each run using this temperature-time history and guess values of the Arrhenius parameters. The sum of the square of the differences between the sugar yields determined by experimentation and by the integration of the rate equations was minimalized by improving the guess values of the kinetic parameters.

The Hooke-Jeeve Pattern Search[6,7,8] was used to improve the guess values of the kinetic parameters and a fourth order Runge-Kutta[9] method was used for the integration of the rate expressions. Details of the methods have been given by Ranganathan.[5]

RESULTS AND DISCUSSION

Cellulose Hydrolysis

Figure 1 shows the typical results for cellulose hydrolysis using 1% sulphuric acid. Experimental and predicted glucose yields are shown for both the isothermal and non-isothermal runs. Temperature profiles are shown by polynomial fits of the experimental temperature data. Similar results were found at 0.5 and 1.5% acid. Table 1 shows the activation energies and pre-exponential factors which gave the best fit of the data for the three acid concentrations. The activation energies, which were independent of acid concentration, were 33 and 21 kcal/g mol for cellulose hydrolysis and glucose degradation, respectively. The pre-exponential factors were power law functions of acid concentration resulting in the following equations for the rate constants for cellulose hydrolysis and glucose degradation, respectively.

$$k_{cellulose} = 6.88 \times 10^{14} [C]^{1.16} \exp(-33{,}000/RT) \quad (6)$$

$$k_{glucose} = 4.29 \times 10^{9} [C]^{0.6} \exp(-21{,}000/RT) \quad (7)$$

where C = concentration of sulphuric acid, weight percent. The activation energy for glucose degradation is lower than the usual value of 30 to 32 kcal/g mol reported in the literature (see Table 2), but agrees with the results of Smith[10] and Grethlein and Converse[11]. The activation energy for cellulose hydrolysis is within the range reported in the literature for other substrates.

Table 1 Kinetic Parameters for Wheat Straw Hydrolysis. Pre-exponential Factors in min^{-1} and Activation Energies in kcal/g mol.

Acid Conc. %	Cellulose A_1 $\times 10^{-14}$	E_1	Glucose A_2 $\times 10^{-9}$	E_2	Hemicellulose A_1 $\times 10^{-20}$	E_1	Xylose A_2 $\times 10^{-15}$	E_2
0.5	3.25	33.0	2.75	21.0	0.70	40.0	0.34	33.8
1.0	6.00	33.0	4.38	21.0	2.90	40.0	1.58	33.8
1.5	12.0	33.0	5.25	21.0	3.60	40.0	2.99	33.8

Hemicellulose Hydrolysis

Typical results for hemicellulose hydrolysis are shown in Figure 2. Hemicellulose requires a lower temperature and shorter time for hydrolysis than does cellulose. The activation energies and pre-exponential factors are shown in Table 1 and the resulting equations for hemicellulose hydrolysis and xylose degradation are as follows:

$$k_{hemicellulose} = 2.25 \times 10^{20} [C]^{1.55} \exp(-40{,}000/RT) \quad (8)$$

$$k_{xylose} = 1.52 \times 10^{15} [C]^{2.0} \exp(-33{,}800/RT) \quad (9)$$

The results compare favourably with the literature values given by Dunlop[12] and Bhandari[13], as shown in Table 3. The hydrolysis results of Mehlberg and Tsao[2] are difficult to compare as they used a more complicated model; however, their activation energy for xylose (monomers) degradation is considerably lower.

Table 2 Literature Values for Hydrolysis of Cellulose for Various Substrates. Pre-exponential Factors are min^{-1} for Unit Acid Concentration and Activation Energies are in kcal/g mol.

Material	Ref.	Cellulose Hydrolysis			Glucose Degradation		
		A	m	E	A	m	E
Glucose	14	-	-	-	$1.85x10^{14}$	1.0	32.7
Glucose	10	-	-	-	$3.84x10^{9}$	0.57	21.0
Glucose	11	-	-	-	$3.96x10^{8}$	0.57	21.0
Corn Stover	13	$2.71x10^{19}$	2.74	45.3	$2.01x10^{14}$	1.86	32.8
Corn Stover	11	$9.62x10^{14}$	1.40	32.8	-	-	-
Solka floc	11	$5.15x10^{16}$	1.14	37.0	-	-	-
Poplar	11	$6.12x10^{15}$	0.99	35.2	-	-	-
White Pine	11	$7.8x10^{13}$	0.96	30.2	-	-	-
Mixed Hardwood	11	$8.98x10^{20}$	1.55	47.1	-	-	-
Newsprint	11	$1.18x10^{17}$	0.70	38.3	-	-	-
Newsprint	15	$2.80x10^{20}$	-	45.1	$4.8x10^{14}$	-	32.8
Douglas Fir	1	$1.73x10^{19}$	1.34	42.9	$2.38x10^{14}$	1.02	32.9
Kraft Paper	6	$2.80x10^{20}$	1.78	45.1	$4.90x10^{14}$	0.55	32.8
Oak Saw Dust	16	$4.40x10^{18}$	1.1	42.9	$2.80x10^{12}$	1.8	30.0

Table 3 Literature Values for Hydrolysis of Hemicellulose for Various Substrates. Pre-exponential Factors are in min^{-1} for Unit Acid Concentration and Activation Energies are in kcal/g mol.

Material	Ref.	Hemicellulose Hydrolysis		Xylose Degradation	
		A	E	A	E
Corn Cob	2				
Xylan I		$1.75x10^{25}$	42.0	-	-
Xylan II		$7.86x10^{17}$	34.0	-	-
Oligomers		$4.31x10^{17}$	31.0	-	-
Monomers		-	-	$1.36x10^{12}$	26.0
Xylose	12	-	-	-	32.0
Corn Stover	13	$3.96x10^{20}$	41.0	$2.65x10^{14}$	32.0

GENERAL DISCUSSION

Preliminary studies using only non-isothermal runs indicated that a unique solution to the kinetic parameters was difficult to obtain as similar yield curves could be obtained using a wide range of activation energies and pre-exponential factors. It was noted, however, that isothermal runs at different temperatures had unique solutions and for this reason, isothermal runs were incorporated into the experimental procedure. Because of this, the low activation energy of 21 kcal/g mol for glucose degradation was found to give a much better fit than the usual literature value of 32 kcal/g mol. More work needs to be done using pure glucose to confirm the proper activation energy for the substance and to determine what effect other chemicals present in the hydrolysates might have on the kinetics. It is also evident from the literature that the kinetic parameters vary with the type of material used, and hence, further work is needed to determine what effect variations in composition, structure, and pretreatment methods have on the reactions.

REFERENCES

1. J.F. Saeman. Ind. Eng. Chem., 37, 42 (1945).

2. R. Mehlberg and P.T. Tsao. "Low liquid hemicellulose hydrolysis of hydrochloric acid", presented at 178th ACS National Meeting, Washington DC, September 1979.

3. H.K. Goering and P.J. Van Soest. Forage Fiber Analysis, Agriculture Handbook, U.S. Dept. Agric., 379(1970).

4. American Association of Cereal Chemists, Technical Committee Report, Cereal Foods World, 26, 295-297 (1981).

5. S. Ranganathan. "Hydrolysis of Wheat Straw using Sulphuric Acid", M.Sc. Thesis (1984).

6. R.D. Fagan, H.E. Grethlein, A.O. Converse, and A. Porteous. Environ. Sci. Technol., 5, 545 (1971).

7. G.S. Beveridge and R.S. Schechter. "Optimization Theory and Practice", McGraw-Hill, New York (1970).

8. D.J. Wilde and C.S. Beightler. "Foundations of Optimization", Prentice-Hall, Englewood Cliffs, NJ (1967).

9. E. Kreyszig. "Advanced Engineering Mathematics", 4th Ed., Wiley, New York (1979).

10. P.C. Smith, H.E. Grethlein and A.O. Converse. Solar Energy, 28, 41 (1982).

11. H.E. Grethlein and A.O. Converse. Proceedings, Royal Society of Canada, Int. Sym. on Ethanol Production, Winnipeg, Canada, 312 (1982).

12. A.P. Dunlop and F.N. Peters. "The Furans", Reinhold Pub. Corp., New York (1953).

13. N. Bhandari, D.G. Macdonald and N.N. Bakhshi. Biotechnol. Bioeng., 26, 320 (1984).

14. H.E. Grethlein. Biotechnol. Bioeng., 20, 503 (1978).

15. G.S. Santini and W.G. Vaux. AIChE Symp. Ser., 72, 99 (1976).

16. J.A. Church and D. Wooldridge. Ind. Eng. Chem. Prod. Res. Dev., 20, 371 (1981).

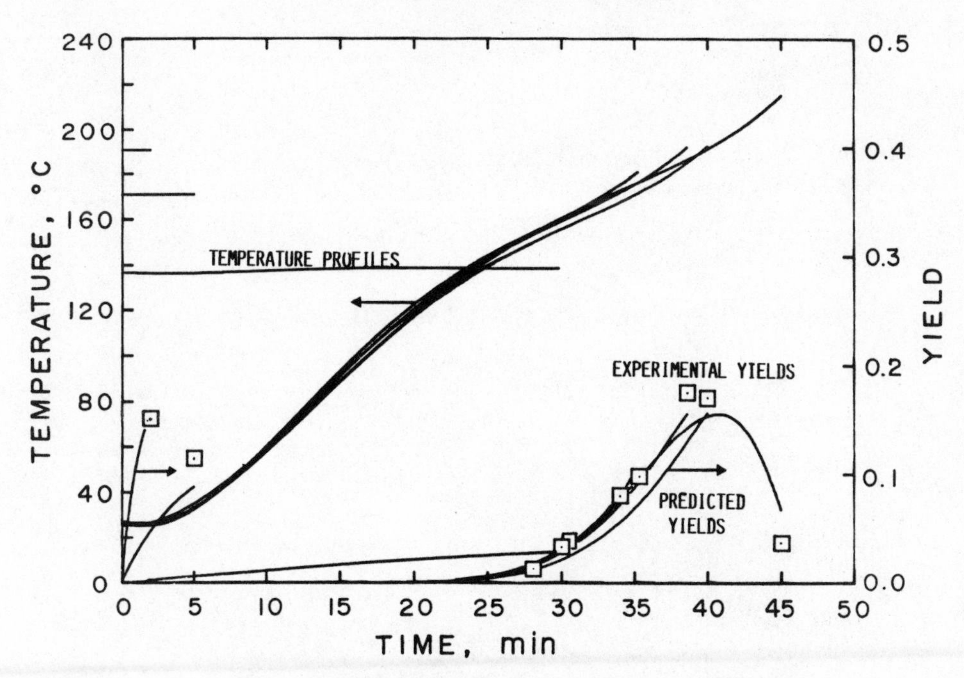

FIGURE 1. CELLULOSE HYDROLYSIS - 1% ACID

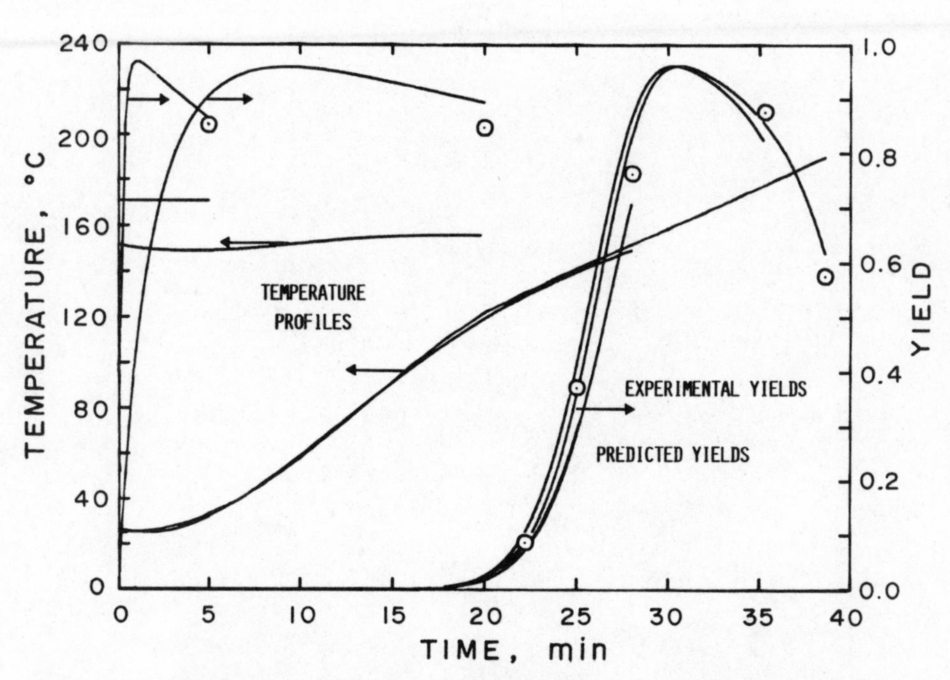

FIGURE 2. HEMICELLULOSE HYDROLYSIS - 1% ACID

LIGNOCELLULOSE DECOMPOSITION BY FUNGI ISOLATED FROM THE FUNGUS GARDEN OF MACROTERMITINAE GROUP OF HIGHER AFRICAN TERMITES

H. Osore

Chemistry-Bioassay Research Unit
ICIPE Research Centre, P.O. Box 30772
Nairobi, Kenya

INTRODUCTION

Biomass in the form of sugars, starch and cellulose is a renewable resource and the annual worldwide production of cellulose alone, anounting to 10^{11} metric tonnes (Blanch and Wilke, 1982) makes it a good candidate for longer range renewable energy needs.

Enzymatic hydrolysis of cellulose into non-petrochemically-based chemicals and fuels has proved to be very attractive because of its specificity and absence of degradation reactions that accompany acid hdrolysis. However, the major impediment to the development of successful bioconversion processes of lignocellulosic materials is the physical protection by lignin of cellulose against cellulolytic enzymes.

Lignin is a particularly recalcitrant substance for enzymatic conversion; consequently, while cellulose decomposition is rather common among microorganisms (Imshenetsky, 1967), lignin decomposition is unusual (Kirk, 1971). More recent explanations of biodelignification studies have involved the use of lignin-degrading organisms (Detroy et al., 1980).

The work described here is part of an on-going study at the ICIPE Research Centre, Kenya, on the possible use of fungi associated with fungus-comb constructing higher African termites for delignification of lignocellulosics, and as a source of cellulolytic enzymes for efficient saccharification of cellulose into fermentable sugars for ethanol production. Fungus combs are deposits of lignocellulosic materials that are continually eaten by the termites and are believed to be involved in the breakdown of lignin (Lee and Wood, 1971).

MATERIALS AND METHODS

The organisms used were cultured from crushed fungus-comb collected from mounds of Macrotermes michaelseni (Osore and Okech, 1983). The fungi of interest that were isolated included, Fusarium semitectum, Trichoderma harzianum, Aspergillus niger, Aspergillus flavus, and Termitomyces spp.

The organisms were grown in the usual liquid media incorporating the following carbon sources individually: Dehydrodivanillyl alcohol, dehydrodivanillin, guaiacyl glycerol β-guaiacyl ether, veratrylglycerol-β-guaiacyl ether, 3-methoxy-4-hydroxy-α-(2-methoxy-phenoxy) β-hydroxypropiophenone,α(-2-methoxyphenoxy) -acetoveratrone and commercial lignocelluloses (Indulin and Polyfon H, from Rimco Trading Corporation, New York). Some monomeric compounds such as p-hydroxybenzoic acid, vanillin and syringic acid were also utilized as substrates.

Cultures were grown in duplicate with continuous shaking throughout the incubation period. In all experiments where lignocelluloses were employed, the substrates were autoclaved dry and then combined aseptically with the sterile mineral salts.

After 15-21 days incubation, the culture filtrates were obtained using a Millipore filtration system. To recover the carbon source and its degradation products, the following procedure was employed:

The culture supernatant was acidified to pH 1.5 with concentrated HCl and then extracted successively with aliquots of ethyl acetate to recover the low molecular weight compounds. The ethyl acetate extractives were combined and anhydrous sodium sulphate added to remove any traces of moisture. The ethyl acetate was removed by evaporation under vacuum after filtering off the sodium sulphate.

Gas Chromatography

Trimethylsilyl (TMS) derivatives of the resulting dry extracts were prepared in dioxane, pyridine or ethyl acetate, according to the method of Lundquist and Kirk (1971).

Gas chromatography was accomplished on a Packard model 428 instrument. The stationary phase was OV-1 and the temperatures were set as follows: injection 285°, detector 230°, and column 220°The carrier gas was nitrogen and detection was by flame ionisation (FID).

High Performance Liquid Chromatography (HPLC)

This was accomplished using a Varian model 5000 liquid chromatograph equipped with a reverse phase, MCH5 micropak column, using an acetonitrile:water:acetic acid (49:49:2) solvent system.

The acidic component of the mixture suppresses the ionisation of phenols. All samples for HPLC analysis were dissolved in acetonitrile.

Sugars were assayed on a micropak-NH2-10 column (Varian) using acetonitrile:water (75:25) as the developing solvent.

Enzyme Assays

(i) Cellulase: For the assay of cellulase activity 1% w/v carboxymethyl cellulose (CMC) was used routinely as substrate. Enzyme activity was determined by measuring the release of reducing sugars in an incubation mixture comprising 1.0 mL CMC, 1.0 mL acetate or phosphate buffer pH 4.6 and 0.2 mL enzyme solution or distilled water (blank).

Both crude enzyme preparations and fractions obtained after purification by ion-exchange chromatography were used in the assays for cellulase activity. Glucose released was estimated by the glucose oxidase method (Osore and Okech, 1983).

(ii) Xylanase Activity: 1% xylan in acetate buffer pH 4.0 was the substrate in an incubation medium consisting of 0.5 mL 1% xylan suspension, in 0-1M sodium acetate buffer pH 4.0, 1 mL of 0.1M sodium acetate and 0.5 mL of the enzyme solution.

The mixture was incubated at 40° for 1 hour and the reaction quenched with 0.2 mL Na_2CO_3 solution, thereby shifting the pH to the alkaline range. Xylose was determined by the glucose oxidase method (Osore and Amoke, 1984).

RESULTS

Table 1 shows the cellulase activity of Termitomyces conidiophores after various purification steps on DEAE-Sepharose, and ammonium sulfate precipitation followed by CM-Sepharose chromatography. A 30-fold purification was achieved.

The fruiting bodies of Termitomyces showed considerably high xylanase activity in crude buffer extracts and after ion-exchange chromatography on DEAE-Sephadex A25 and gel-filtration on Biogel P100 (Table 2). One hundred-fold purification was achieved after gel-filtration on Biogel P100.

Fig. 1 shows an HPLC analysis trace of dimeric compounds structurally related to subunits within the lignin biopolymer. These compounds were synthesized for use in biodegradation studies in our laboratory. Fig. 2 shows that monomeric compounds such as protocatechuic acid, syringic acid, vanillin and trimethoxybenzoic acid, which are some of the single-ring compounds found in culture fluids of fungi growing on lignocellulose as carbon source.

Fusarium, Trichoderma and Aspergillus niger metabolized 3-methoxy-4-hydroxy-β-hydroxypropiophenone in three distinct patterns as shown in the HPLC traces in Fig. 3. Fifteen day old cultures of Aspergillus niger metabolized α(2-methoxyphenoxy) β-hydroxypropioveratrone into a variety of low molecular weight compounds that gave multiple G.C. peaks after silylation (Fig. 4). Complex lignocellulosic materials such as Indulin, were also partially degraded by Aspergillus niger into low molecular weight intermediates that could be separated by HPLC (Fig. 5).

DISCUSSION

Results of the studies presented above demonstrate that conidiophores (conidia) and fruiting bodies (mushrooms) of the fungus Termitomyces have substantial cellulase and xylanase activities, and that this fungus together with others associated with it in the termite mounds (Fusarium semitectum, Aspergillus niger and Trichoderma harzianum) are capable of degrading a variety of lignin model compounds and lignocelluloses.

If Termitomyces can be propagated on a large scale, it would appear to offer a cheap and useful source of cellulase and xylanase enzymes for efficient saccharification of lignocelluloses into sugars for the production of liquid fuels and chemicals. Pentose sugars derived from the action of xylanase have recently been shown to be amenable to fermentation to ethanol by certain types of yeasts (Flickinger, 1980). This should accelerate the search for microbial sources of xylanase enzymes. Indeed, pentose sugars derived enzymatically from the hitherto unutilized hemicellulose xylan, may very well prove to be the least expensive fermentable carbohydrate available for conversion to liquid fuels (Osore and Amoke, 1984).

Our results with model compounds employing HPLC and G.C. techniques show that, when coupled with mass spectrometry, the structures of low molecular weight degradation products can be easily determined, and hence the pathways of degradation for individual compounds by specific organisms.

We are currently working on the characterization of ligninolytic and cellulolytic enzymes involved in the degradation processes. A number of Actinomyces species associated with the fungus-garden have also been isolated and are being studied for their abilities to decompose ^{14}C-labelled substrates.

ACKNOWLEDGEMENTS

This investigation received financial support from the Italian Government through grant no. TC/INT/5/084 offered jointly to the ICIPE Research Centre and the International Atomic Energy Agency, Vienna.

I am thankful to the Director at ICIPE for continuous encouragement and to M.A. Okech for the supply of microorganisms; to Mr. P.O. Amoke and M. Kotengo for technical assistance. Last but not least, I am grateful to the Lord for revealing some of the subtleties of his creation.

REFERENCES

Blanch, H.W. and C.R. Wilke (1982). Sugars and chemicals from cellulose. Rev. Chem. Eng., 1, 71-119.

Detroy, R.W., L.A. Lindenfelser, G. St. Julian Jr., and W.L. Orton (1982). Saccharification of wheat-straw cellulose by enzymatic hydrolysis following fermentative and chemical pretreatment. Biotechnology and Bioengineering Symp., 10, 135-148.

Flickinger, M.C. (1980). Current biological research in conversion of cellulosic carbohydrates into liquid fuels. How far have we come? Biotechnology and Bioengineering, 22, Suppl. 1, 27-48.

Imshenetsky, A.A. (1967). Decomposition of cellulose in the soil. In T.R.G. Gray and D. Parkinson (eds.). The Ecology of Soil Bacteria, Liverpool University Press, 256-269.

Kirk, T.K. (1971). Effects of microorganisms on lignin. Ann. Rev. Phytopathol., 9, 185-210.

Lee, K.E. and T.G. Wood (1971). Termites and soils, Academic Press, London, 17.

Lundquist, K. and T.K. Kirk (1971). Acid degradation of lignin IV. Analysis of lignin acidolysis products by gas chromatography using trimethylsilyl derivatives. Acta Chem. Scan., 25, 889-894.

Osore, H. and M.A. Okech (1983). The partial purification and some properties of cellulase and β-glucosidase of Termitomyces conidiophores and fruit bodies. J. Appl. Biochem. (Biotechnology), 5, 172-179.

Osore, H. and P.O. Amoke (1984). Partial purification and properties of extracellular xylanase from Termitomyces spp., J. Appl. Environ. Microbiol., submitted November, 1984.

Table 1. Purification of Cellulase from Buffer Extracts of Termitomyces Conidia

Enzyme Sample Purification Steps	Total (mg)	Recovery (%)	Specific Activity (nmol glucose mg protein^{-1} h^{-1})	Total Activity Purification (fold)
Crude extract (100 mL)	350	100.0	0.99	1.0
DEAE-Sepharose (100 mL)	6.0	1.7	22.75	23.0
Precipitation with saturated $(NH_4)_2SO_4$ then chromatography on CM-Sepharose	4.0	1.14	29.75	30.0

Table 2. Purification of Xylanase from Buffer Extracts Termitomyces Fruit Bodies

Enzyme Sample Purification Steps	Total Protein (mg)	Recovery (%)	Specific Activity nmol xylose mg pr^{-1} h^{-1}	Total Activity Purification (fold)
Crude extract (50 mL)	242.5	100	0.896	1
DEAE-Sephadex A25	0.515	0.21	5.68	6.33
Biogel-P100	0.101	0.042	100.0	111.6

1. 3 — methoxy — 4 — hydroxy — (2 — methoxyphenoxy) B — hydroxypropiophenone
2. 3, 4. dimethoxy compound of 1.
3. Dehydrodiisoeugenol
4. Dihydrodehydro diisoeugenol methyl ether.
5. O, O' biphenyldehydrodimer of propioguaiacone

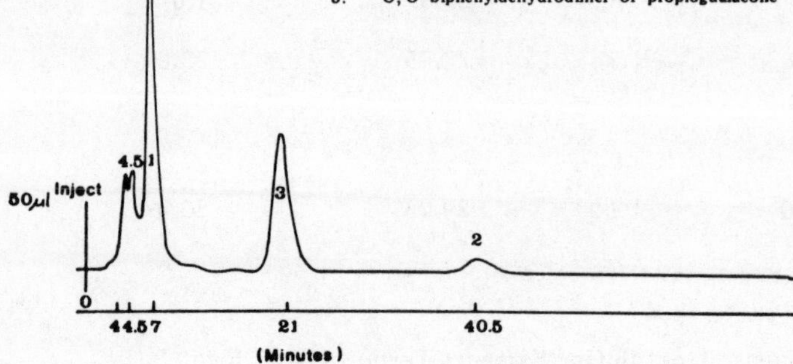

Fig. 1.HPLC analysis of dimeric lignin model compounds using an Acetonitrile: Water: Acetic acid (49:49:2) solvent system.

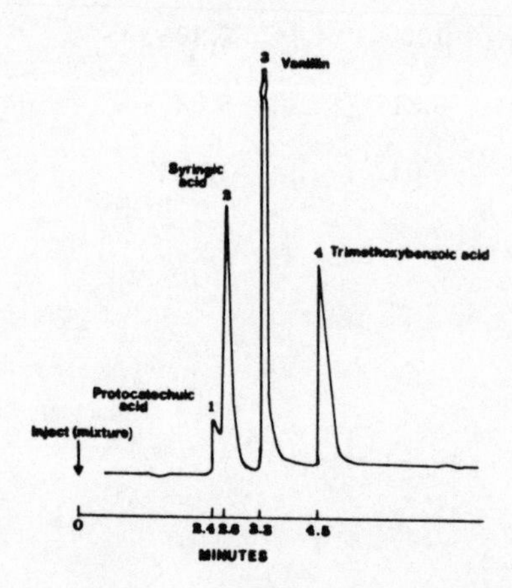

COLUMN : MCH—5, Reverse Phase, 4mm × 30 cm.
ELUENT : Acetonitrile / Water / Acetic acid [49:49:2]
DETECTION : UV 254 nm.

Fig. 2. HPLC separation of single ring degradation products of lignin.

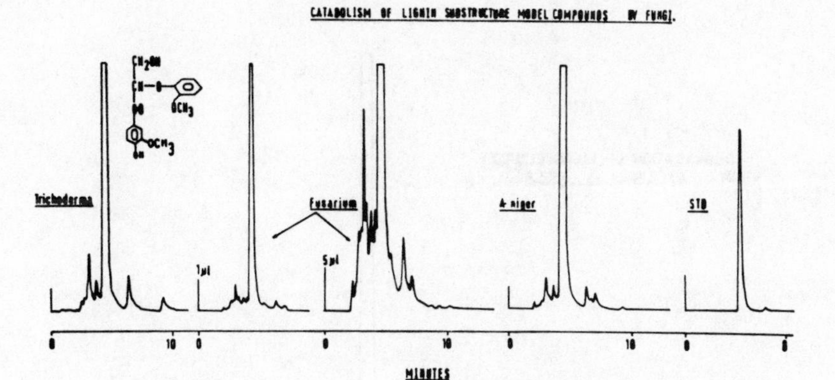

Fig. 3. HPLC analysis of the pattern of degradation of 3-methoxy-4-hydroxy- β-hydroxypropiophenone by Trichoderma harzianum, Fusarium semitectum and Aspergillus niger.

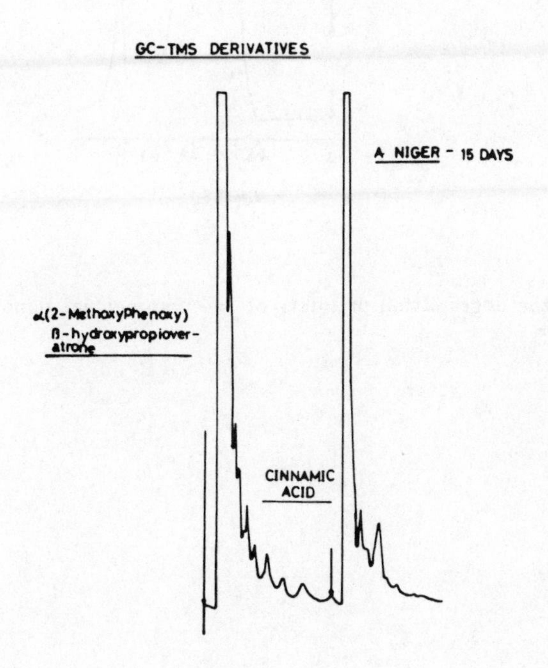

Fig. 4. GC analysis of the pattern of degradation of α(2-methoxyphenoxy)- β-hydroxy-propioveratrone by Aspergillus niger.

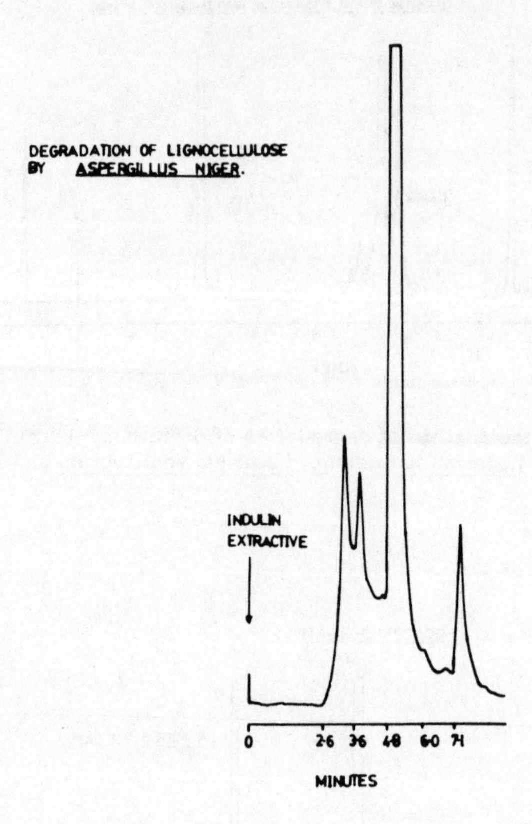

Fig. 5. GC analysis of the degradation products of the commercial lignocellulose indulin by Aspergillus niger

ANHYDROUS HYDROFLUORIC ACID SOLVOLYSIS OF CELLULOSE

C.M. Ostrovski and J. Aitken

Canertech Inc., University of Manitoba,
Winnipeg, Manitoba,
R3B 1N3 Canada

ABSTRACT

Anhydrous hydrofluoric acid (HF) in the gaseous phase was utilized to hydrolyze poplar wood to produce monomeric sugars for fermentation to fuel ethanol. Test results have shown that sugar yields for hexose and pentose consistently exceeded those by other hydrolysis techniques. Conversion yields of cellulose to glucose greater than 90% of the theoretical, and of hemicellulose to xylose in excess of 80% have been achieved. Since HF is being used in the vapour phase, it has been possible to hydrolyze the wood and then desorb and recycle the HF to levels exceeding 99% (by weight) of the HF used in the hydrolysis. Preliminary studies on the fermentation of hexose and pentose sugars indicate that it may be possible to produce between 300-400 litres of ethanol from one metric ton of oven dry poplar.

INTRODUCTION

The Ethanol-from-Cellulose Program began in late 1981, when Canertech was designated lead agency for Canada for the development of a pilot plant and related research and development program. Initially, the Program focussed on the use of poplar as the feedstock. This hardwood is found throughout Canada. Poplar grows well on marginal land, has a history of being under-utilized, and has lent itself remarkably well to hybrid development. These aspects favour poplar for silviculture which would not be in conflict with agricultural lands used for food production.

Historically, ethanol production technology has been developed using softwoods and it was hoped that the best of the unit processes for the current technology could be adapted for use into an integrated process, whose end product would be ethanol at a cost competitive with gasoline. Such an integrated process must (1) separate the cellulose and hemicellulose fractions from the lignin, (2) convert the cellulose and hemi-cellulose to monomeric sugars and (3) ferment the monomeric sugars to ethanol. The first step can be performed physically and/or chemically, the second step requires a hydrolyzing agent and the third can be carried out by either fungi or bacteria. In integrated processes for the production of ethanol from softwoods, the hydrolysis step is carried out chemically using a weak acid, and the fermentation step acts only on the hexose component of the hydrolyzate. These processes also represent a possible route to convert hardwoods to ethanol. However, other possible hydrolysis methods include:

- use of strong acids
- use of organic solvents followed by a weak acid treatment
- microbiological means or enzymatic hydrolysis.

Due to structural and chemical differences between softwoods and hardwoods, processes applicable to one species are not necessarily applicable to the other, particularly with reference to processing technology and/or economics. Two principal characteristics of hardwoods that affect the process type are:

- the hardwood lignin content (approximately 22% of the oven dry weight of the wood) is less than that for softwoods, which could lead to a more rapid separation of the lignin, hemi-cellulose and cellulose, and

- hardwoods contain more pentose polysaccharides than softwoods (approximately 20% of the oven dry weight of the wood), and these represent more than one quarter of the saccharides found in poplar. Therefore, for sound economics, pentose fermentation is necessary. A summarized comparison of the hardwood Populus tremuloides, and softwoods in general is presented in Table 1.

Table 1. Composition of intact wood, as weight per cent of oven dry

		Populus tremuloides (Hardwood)	Soft Woods
Lignin		21-22	27-28
	Hexoses		
Sugars	Glucose	50-52	50
	Mannose	1.5-2.3	10
	Galactose	0.7	4.5
	Pentoses		
	Xylose	20-21	6-7
	Arabinose	0.3	2-3

THE HF HYDROLYSIS PROCESS

Since at the outset of the program, Canertech did not own any technology, and objective and independent evaluation was carried out to assess available process technologies and their unit processes, including hydrolysis processes, for their performance with respect to the economic production of fuel ethanol from Canadian poplar. As a result, Canertech undertook to concentrate and develop, as a basic unit process for the integrated process, the most attractive hydrolysis process based on the use of anhydrous hydrofluoric acid (HF) in the vapour phase, despite the more rudimentary status of its technology.

The hydrogen fluoride hydrolysis unit process when fully developed will then be combined with the other unit operations to form an integrated ethanol process. The integrated ethanol pilot plant shown schematically in Figure 1, is based on one metric tonne of oven dry wood. Theoretically, this tonne will yield 438 litres of ethanol from the hexose and pentose sugars produced.

The use of anhydrous hydrogen fluoride for hydrolysis of wood was initially employed in Germany, where in 1938 a pilot plant was built for this purpose. However, the work was abandoned during the war and only recently revived by, for example, the following groups: Michigan State University (United States) (2); Biotechnisk Institute (Denmark) (3); Hoeschst AG (Germany) (4); Forintek, under contract to the National Research Council of Canada (5).

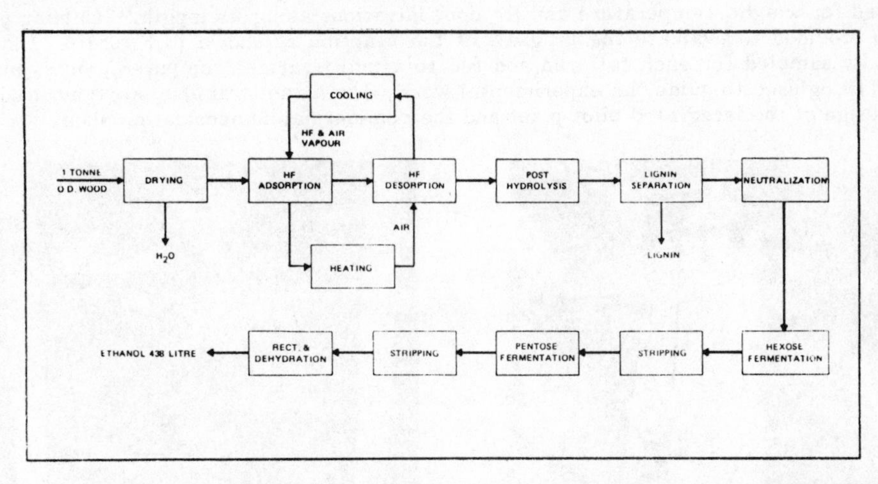

Figure 1: Integrated HF Hydrolysis of Wood to Ethanol Process (1)

The information provided by these laboratories has strongly influenced the design of the ethanol from cellulose test facility, shown in Figure 2, which is located at the University of Manitoba.

Figure 2: General View of the Two-Storey Test Facility Including Laboratory and Control Room on the Left

HF HYDROLYSIS TEST FACILITY

The heart of the test facility is the reactor, Figure 3, whose working capacity is 20 kg oven dry weight of poplar chips. Up to March 1984, tests had been carried out using batches simulating continuous feed. Recent modifications to the facility allow for testing in continuous or semi-continuous modes. The reactor is suspended from a load cell, and monitored for weight, temperature and HF concentrations along its length. Sampling ports are also provided to evaluate the progress of the reaction as shown in Figure 4. Data is continually sampled for each test run and fed to a multi-variant computer process model which is being used to guide the experimental work. The model will also prove most useful in the design of the integrated pilot plant and the commercial demonstration plant.

Figure 3: The Reactor, Showing Flexible Outlet Connection, HF Concentration Sampling System (Syringes and Lines)

Figure 4: Second Floor View of the Reactor Including Suspension and Loadcell Mechanism

Poplar used in the test runs was obtained from local stands. Logs were debarked and chipped in a standard 2.5 cm chipper and then screened through a 2.5 cm mesh screen. The fresh chips contain about 50% moisture by weight and must be dried prior to any exposure to HF. Once the chips are adjusted to the proper moisture content, a charge of chips is top loaded into the reactor.

HF used in the process is stored as a liquid in pressure vessels as shown in Figure 5. Vaporized HF is fed into the reactor where it reacts with the chips, and after the appropriate contact time, the HF is then desorbed from the chips using hot air. The results obtained show that over 99% by weight of HF is recovered from the chips. The recovered HF is recycled in the process.

As a result of HF solvolysis (part of the solvolysis reaction is shown in Figure 6), a hydrolyzate is produced containing short chain oligosaccharides that have a degree of polymerization mostly in the range 2 to 6. A visual comparison of the hydrolyzed chips and untreated poplar chips is shown in Figure 7. The penetration of HF in the wood is shown in the sequential photographs in Figure 8a, b, c. The hydrolyzate requires a mild secondary hydrolysis treatment to produce monomeric sugars, principally glucose and xylose. This post hydrolysis was carried out as part of a comprehensive investigation by Atomic Energy of Canada (WNRE), Pinawa (Manitoba) under contract to the Ethanol-from-Cellulose Program.

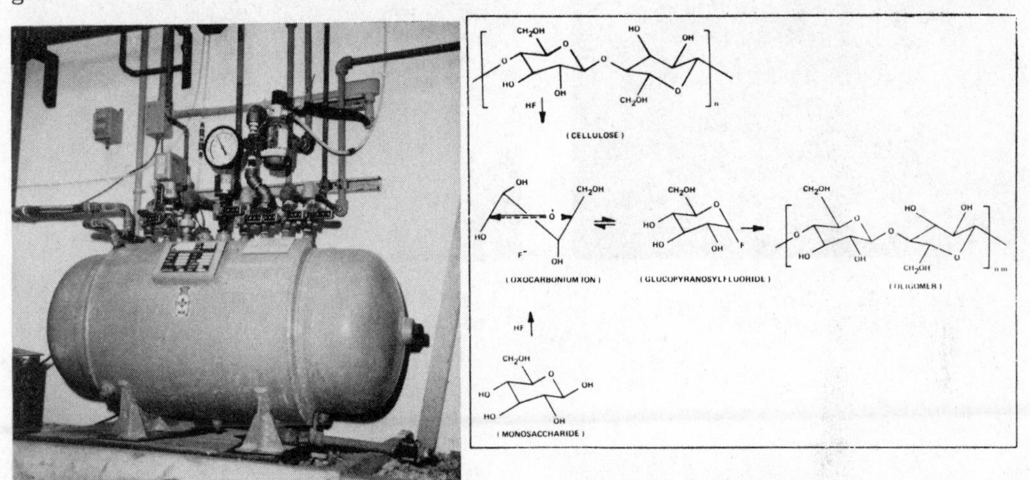

Figure 5: HF Supply Tank, Weigh Scale Mounted, Pressure Rated to 300 psi

Figure 6: HF Solvolysis of Cellulose

Figure 7: Comparison of Virgin Poplar Chips and HF Hydrolyzed Chips

Figure 8a: Cross Section of Untreated Wood Chip

Figure 8b: Cross Section of Wood Chip Partially Treated with HF

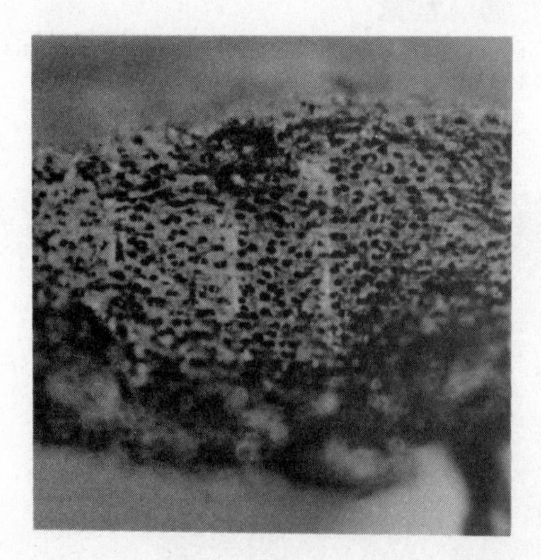

Figure 8c: Cross Section of Wood Chip After Full Treatment with HF

Sugar analysis results show that conversions in excess of 90% of the theoretical in the case of glucose and over 80% in the case of xylose are obtainable and reproduceable. Some tests have resulted in 94% glucose and 87% xylose conversions.

In the light of available and developing fermentation technologies, the Ethanol from Cellulose Program has, as an objective, the production of high sugar content solutions. In the optimization runs, it has been relatively easy to obtain sugar solutions to 15% w/v and it appears very likely that this concentration could be increased to the 20 to 30% w/v range. To take advantage of these concentrated solutions, direct fermentation of the hydrolyzate is required. It was thus important to investigate whether new and currently used fermentation microorganisms were capable of acting on this hydrolyzate.

FERMENTATION

Fermentation studies have been continuing in parallel with the hydrogen fluoride hydrolysis tests. These are being conducted by the National Research Council (Ottawa). Since pentose sugars represent a high proportion of the total sugars, one line of investigation has given special emphasis to pentose sugar fermentation to ethanol. Since the start of the program, about 500 strains of microorganisms have been evaluated, of which seven have been selected for further intensive studies. Some of these strains have given very encouraging results. Using either pure xylose or the HF hydrolyzate, particular organisms have produced 5% ethanol in less than 20 hours. The other line of investigation has followed the traditional path exploring the use of different microorganisms for hexose conversion to ethanol. Fermentation studies have shown good conversion efficiencies for both glucose and xylose by a mutant of *Clostridium saccharolyticum* (6). This improved conversion was accompanied by the organism's very significant increase in its tolerance to ethanol. Both fungi and bacteria are being used in these investigations, and the results have been very encouraging for both pentose and hexose fermentation.

ENGINEERING AND SAFETY

From an engineering standpoint, hydrogen fluoride has proved to be a challenging material to work with since there is only limited experience in its use. It is a gas at room temperature and highly toxic, but its presence is immediately detected visually or by smell. Once proper precautions have been taken and the personnel trained in working with the material, it is straightforward to use. The test facility incorporated a variety of construction materials, instruments and systems used in the everyday operation of the facility, in order that these may be evaluated under operational conditions. Little, if any, corrosion was detected in the materials used. In fact, certain materials generally not recommended for use with hydrogen fluoride have actually proved satisfactory, at least under test conditions.

Since the facility was located on a University campus, extra precautions were taken to ensure that no pollution problems arose, particularly air pollution. Scrubbers and filters were installed to control the facility exhausts. These have functioned very well and no detectable levels of hydrogen fluoride have been found during the many months of operation.

CONCLUSION

Hydrogen fluoride hydrolysis continues to demonstrate its merits, not only in the high conversion of cellulose to sugars, but also in its relative ease of handling and in the very important requirement for high efficiency in the recovery and reuse operations. In addition, studies are showing that the wood-sugar solution, as produced by the hydrogen fluoride hydrolysis of *Populus tremuloides*, has the potential of being a good substrate for industrial ethanol fermentation.

REFERENCES

1. Canertech Inc., Ethanol-from-Cellulose Program "Assessment of Hydrogen Fluoride Wood Hydrolysis Process", Internal report (confidential) prepared by W.L. Wardrop and Associates Ltd., Engineering Consultants, Winnipeg, Canada, April 1983.

2. Lamport, D.T.A., Hardt, H., Smith, G.; Mohrlok, S., Hawley, M.C., Chapman, R. and Selke, S.: 1981. Energy from Biomass, W. Palz, P. Chartier and D.O. Hall (eds.), (London) Applied Science Publishers, pp 292-297.

3. Defaye, J., Gadelle, A. and Petersen, C.: 1981. Energy from Biomass, W. Palz, P. Chartier and D.O. Hall (editors), (London) Applied Science Publishers, pp. 319-323. Also private communication from T. Bentsen.

4. Private communication from H.-M. Deger.

5. Private communication from L. van den Berg (N.R.C.) and J.N. Saddler (Forintek).

6. Murray, W.D., Wemyss, K.B., Khan, A.W.: 1983. "Increased Ethanol Production and Tolerance by a Pyruvate-Negative Mutant of Clostridium saccharolyticum", National Research Council of Canada, No. 21316.

Section 2

Production of Fuels and Solvents

EVALUATION OF TECHNOLOGIES FOR CONVERSION OF BIOMASS SUGARS TO ETHANOL AT PILOT SCALE USING ZYMOMONAS MOBILIS

J. Fein, R. Charley, R. Droniuk, D. Good,
K. Hopkins, G.R. Lawford and B. Zawadzki

Weston Research Centre, Toronto, Ontario M4W 2L3
and H. Lawford
Dept. of Biochemistry, University of Toronto, Toronto, Ontario M5S 1A8

ABSTRACT

A state of the art pilot plant has been designed for the continuous conversion of 1 oven dried ton of wood per day. It has been equipped for process optimization and data acquistion in the demonstration of the Bio-hol process (European patent 0047641). Previous studies by our group and others have shown that the microorganism, Zymomonas mobilis, has considerable advantages over yeast. Better product yield and specific and volumetric productivity have been demonstrated for laboratory-scale continuous processes. These advantages have been confirmed at pilot-scale using several process configurations. A continuous process has been developed at the pilot plant for producing inexpensive sugar hydrolysates from wood and agricultural residues. Data will be presented on the fermentation performance of both Z. mobilis, and Saccharomyces cerevisiae on this feedstock, and on current efforts for increasing productivity through feedstock refining and strain optimization.

(The Bio-hol Developments venture has had financial support through the Ontario Ministry of Energy and Energy, Mines and Resources, Canada (CREDA).)

INTRODUCTION

Interest in the production of fermentation ethanol for use as a liquid fuel has increased within the past several years because of rising fuel costs and the wish by western governments to reduce or restrict their dependence on imported oils. Economics of such processes will be largely dependent on the ability to provide inexpensive, locally grown renewable fermentation feedstocks on a year-round basis. In conventional fermentations with molasses or corn starch hydrolysates, the substrate cost is approximately 60% of the overall production expenses (1).

The process presented here provides a series of novel technological developments for the conversion of inexpensive lignocellulosic materials to fuel-grade ethanol. A state-of-the-art pilot plant designed for the conversion of 1 oven dried ton of wood per day has been commissioned and operated on behalf of Bio-hol Developments Ltd., a joint venture between George Weston Ltd. and the Ontario Energy Corporation.

RESULTS AND DISCUSSION

Lignocellulose Pretreatment and Hydrolysis

A number of design criteria were established to evaluate technologies for converting lignocellulose into fermentation substrates. Ideally the system should be capable of both pretreatment and hydrolysis to permit maximum downstream flexibility; the system should be continuous, energy efficient and capable of controlled, safe operation. For these reasons extrusion technology was selected and, following preliminary studies at Wenger Mfg., Sabetha, Kansas, a custom-designed extruder was installed at the pilot plant facility.

In the extruder, lignocellulose materials are subjected to high shear forces at elevated temperatures and pressures, rupturing the particles and increasing their susceptibility to hydrolytic activity. Simultaneous pretreatment and hydrolysis accomplished by injection of acid has produced glucose yields of approximately 60% of the theoretical maximum with poplar and pine wood chips.

Fermentation Technology

Continuous fermentation using the bacterium Zymomonas mobilis has been shown (2-4) to exhibit a number of significant advantages over the use of Saccharomyces cerevisiae (Table 1). Typically Zymomonas mobilis, compared to yeast, produces ethanol with a 5% increase in substrate conversion efficiency, and at a much faster rate (specific rate of product formation, Qp). Also, Z. mobilis is capable of high rates of ethanol formation under conditions of restricted (ie. uncoupled) growth.

The higher substrate conversion efficiency is particularly important since it would result in an estimated 2-3% reduction in ethanol production costs for commercial operation (5). The increased specific rate of ethanol production would allow a higher hydraulic throughput, and hence reduced plant size and capital investment, compared to a yeast process with the same cell concentration. The ability to produce ethanol while growth restricted has been exploited in a patented two-stage process, allowing a high final product concentration (100 g/L) to be achieved. Also, product recovery costs and the potential for contamination were reduced (2).

Process Scale Up

To determine whether the laboratory-demonstrated advantages could be achieved during prolonged operation at pilot scale, fermentation research at the pilot plant has concentrated on the evaluation of a number of process designs using Z. mobilis ATCC 29191.

The fermentation equipment at the pilot plant was designed for flexible operation in a number of configurations. Single and 2 stage continuous culture, both with and without cell recycle, at a capacity of 3.0 m^3/d were among those configurations tried. Using commercial corn starch hydrolysate as substrate, the various process designs were successfully operated over long periods at pilot scale. The data compared very favorably with similar yeast-based processes (Table 2). Most significantly, the high substrate conversion efficiency (previously only demonstrated for Z. mobilis in the laboratory environment) was achieved routinely.

Wood Hydrolysate Fermentation

It has long been recognized that unrefined acid wood hydrolysates contain a variety of toxic substances that are inhibitory both for growth and fermentative activity of yeasts (8,9). These substances include furfural, hydroxymethylfurfural. various organic acids, lignin monomers and lignin degradation products, and heavy metal ions resulting from equipment corrosion. Many of these substances are generated under the rigorous conditions required to bring about crystalline cellulose hydrolysis. In addition, the hydrolysates may contain a high salt content resulting from the acid-neutralization.

Hydrolysates prepared from wood chips and agricultural residues have only recently been available for fermentation studies. The suitability of the hydrolysates produced as substrates for ethanol production has been assessed using a broth dilution MIC test (10) with S. cerevisiae NCYC 431 and Z. mobilis strains ATCC 29191 and CP4. As expected, both microorganisms experienced growth inhibition at low hydrolysate concentrations (Table 3). Both Z. mobilis strains grew with disturbed (filamentous) morphology at relatively low hydrolysate concentrations.

TABLE 1 Comparison of Yeast and Bacterial Process Parameters in Laboratory Scale Single Stage Continuous Fermentations with and without Cell Recycle

Parameter	Steady State Values at Dmax[a]			
	CSTR - Recycle		CSTR + Recycle	
	SC	ZM	SC	ZM
Glucose feed, (g/L)	100	100	100	100
Dmax, (h^{-1})	0.17	0.23	0.67	3.41
Ethanol conc., (g/L)	41.2	42.5	43.0	44.3
Product yield, Yp/s, (g/g)	0.43	0.44-0.46	0.43	0.46-0.48
Cell density, (g/L)	13.0	3.3	50.0	63.7
Specific productivity, Qp, (g/g/h)	0.54	2.90	0.58	2.37
Volumetric productivity, (g/L/h)	7.0	9.6	29	151
Data Source	12	2	13	This study

[a]Dmax, maximum dilution rate at which the residual glucose conc. = 4 g/L.

abbreviations: SC, S. cerevisiae; ZM, Z. mobilis

Table 2: Comparison of continuous pilot scale fermentation using yeast and zymomonas

Kinetic Parameter	Z. mobilis continuous systems (Bio-hol Developments) Single-Stage	Two-Stage	Pilot-scale yeast continuous ethanol production Alcon Biotechnology[1] Molasses	Cane Juice	Hoehst/ Uhde[2]	National Distillers[3]
Product alcohol concentration (gL^{-1})	53-60	96-105	57	65	80	90-100
Process yield[4]	93-97	88-98	88-91	91	90	88
Productivity ($gL^{-1}h^{-1}$)	9-13	4-5	8	9	16	N/A

Notes:
1 - Alcon Biotechnology process: single-stage continuous culture (stirred tank reactor with yeast recycle by external sedimentation. Data taken from Alcon Biotechnology Ltd. leaflet "Tropical demonstration of the Alcon process."
2 - Hoechst/Unde process: single stage continuous culture (air lift tower reactor) with yeast recycle by external sedimentation (6).
3 - National Distillers and Chemical Corp. process: two-stage continuous culture (stirred tank reactors) with cell recycle by centrifugation (7).
4 - Yield based on total sugar input.
N/A - Not available

TABLE 3 Comparative Growth of Z. mobilis and S. cerevisiae on Unrefined and Refined Wood Hydrolysates

PRETREATMENT SCHEME	MIC'S[1] Z. mobilis ATCC 29191	CP4	S. cerevisiae NCYC 431	MFC[2] Z. mobilis ATCC 29191	CP4
Unrefined	0.8	0.8	1.5	-	-
Partially refined	3.4	3.4	6.8	1.7	1.7
Refined	16.9	16.9	>16.9	4.2	8.4

1. Determined by the broth dilution method (10) in complex medium (glucose, yeast extract plus salts (3)). Hydrolysate concentrations were based on their initial glucose concentration (%w/v).

2. MFC, minimal substrate concentration giving rise to filamentous growth of the bacteria.

Wood Hydrolysate Fermentation

To investigate this problem of toxicity further, MIC's were determined for Z. mobilis ATCC 29191 for several potential inhibitors frequently present in wood hydrolysates, and for three salts. As shown in Table 4, all of the test compounds could bring about growth inhibition of the bacteria. At sub-growth inhibitory concentrations, most of the agents also caused gross morphological disturbances (i.e. filament formation).

Since the success of the Bio-hol process will ultimately depend on the ability of the microorganism to utilize the cellulosic hydrolysate, the problem of toxicity of the substrate has been attacked simultaneously from two different directions.

1. Substrate clean-up: purification by chemical and physical treatments has been successful for selective removal or destruction of inhibitory substances.

2. An on-going long term mutagenesis and selection programme, for the isolation of more growth-tolerant strains.

Using a proprietary sequence of clean-up steps, growth on, and fermentation of, wood hydrolysate by S. cerevisiae ATCC NCYC 431 and Z. mobilis strains ATCC 29191 and CP4 have been improved significantly. The data presented in Table 3 show the effect of substrate clean-up on the growth of these microorganisms and on the cellular morphology of Z. mobilis. Throughout the study, the yeast has shown greater tolerance to impure substrates than either of the bacterial strains evaluated, and Z. mobilis strain CP4 has been less inhibited and less subject to morphological disturbance than strain ATCC 29191.

Fermentation by Z. mobilis CP4 of partially refined wood hydrolysate stream produced a high yield of ethanol. This was a preliminary batch fermentation and the yield was based on the wood hydrolysate glucose supplied (94% theoretical maximum). Unfortunately, a long lag period prior to hydrolysate utilization resulted in a low overall productivity, reflecting the requirement for further clean-up. It is possible that growth inhibitory hydrolysates could be used to exploit the ability of Z. mobilis to continue high rate fermentation while growth restricted either by use of a two-stage process or by feeding a less refined hydrolysate at a growth limited rate.

CONCLUSIONS

1. A continuous process has been developed for the production of cellulosic hydrolysates (60% yield).

2. Strain improvement and hydrolysate purification are required for good fermentability of this material.

3. The advantages of Z. mobilis documented in laboratory studies (Yp/s, Qp) have been confirmed at pilot scale.

4. Preliminary studies have shown that high glucose to ethanol conversion yields are attainable with refined hydrolysates.

ACKNOWLEDGEMENTS

The authors thank D. Potts, R. Edamura, A. Effio, P. McLaren, and D. Docherty for provision of the wood hydrolysates and for the analytical support.

Table 4: Effects of potential wood hydrolysate inhibitors and salts on the growth and morphology of Z. mobilis ATCC 29191

Factor	Minimum[1] Inhibitory Concentration (% w/v)	Morphological disturbance (causative concentration) (% w/v)
Acetic acid	1.72	Short-medium length filaments (1.23-1.27)
Formic acid	1.57	Short filaments (1.23)
Propionic acid	1.37	Plump cells, chains (0.98-1.18)
Levulinic acid	4.90	Plump cells (3.43-3.92)
Hydroxymethyl-furfural	1.47	Short-medium filaments (0.49-0.98)
Furfural	0.22	Short filaments (0.147-0.197)
Phenol	0.29	Medium-long filaments (0.049-0.88)
KCL	1.96	Filaments and club-shaped cells (0.98)
NH_4Cl	1.96	Bulging cells (0.98) long filaments (0.49-0.8)
$(NH_4)_2SO_4$	2.94	Club-shaped cells (1.5-2.0) filaments (1.0)

[1] determined by the broth dilution method in complex medium (glucose, yeast extract plus salts). For details, see ref. 11.

REFERENCES

1. Righelato, R.C. (1980) Phil. Trans. T. Soc. Lond. B. 290, 25-34.

2. Lawford, G.R., B.H. Lavers, D. Good, R. Charley, J. Fein and H.G. Lawford (1983) In: Proceedings of the Royal Society of Canada International Symposium on Ethanol from Biomass, 482-507, Eds. H.E. Duckworth and E.A. Thompson, Royal Society of Canada (Ottawa).

3. Fein, J.E., H.G. Lawford, G.R. Lawford, B.C. Zawadzki and R.C. Charley (1983) Biotechnol. Lett. 5, 19-24.

4. Rogers, P.L., K.J. Lee, M.L. Skotnicki and D.E. Tribe (1982) In: Advances in Biochemical Engineering, Vol. 2, 37-84, Ed. A. Feichter, Springer-Verlag (New York).

5. Flannery, R.J. and A. Steinschneider (1983) Biotechnology 1, 773-776.

6. Faust, U., P. Prave and M. Shlingmann (1983) Process Biochem. 18, (3) 31-37.

7. Muller, W.C. and F.D. Miller (1983) U.S. Patent No. 4,385,118.

8. Leonard, R.H. and G.H. Hajny (1945) Ind. Eng. Chem. 37, 390-395.

9. Fein, J.E., S.R. Tallim and G.R. Lawford (1984) Can. J. Microbiol. 30, 682-690.

10. Anderson, T.G. (1970) In: Manual of Clinical Microbiology, 299-310, Eds. J.E. Blair, E.H. Lennette and J.P. Truant, Amer. Soc. Microbiol, (Baltimore).

11. Fein, J.E., D.L. Barber, R.C. Charley, T.J. Beveridge and H.G. Lawford (1984), Biotechnol. Lett. 6, 123-128.

12. Cysewski, G.R. and C.R. Wilke (1976) Biotechnol. Bioeng. 18, 1297-1313.

13. Cysewski, G.R. and C.R. Wilke (1977) Biotechnol. Bioeng. 19, 1125-1143.

AN OVERVIEW OF THE ENVIRONMENTAL IMPACTS ANTICIPATED FROM LARGE SCALE BIOMASS/ENERGY SYSTEMS

T.C. McIntyre
Science Policy Analyst, Policy & Expenditure Management Branch
Environmental Strategies Directorate
Environmental Protection Service, Environment Canada
Hull, Quebec

INTRODUCTION

In recognizing our obligations in the energy arena in general and alternative energy in particular, Environment Canada has developed an energy strategy based partly on the recognition that:

> "A policy for the development of energy resources must be created in the context of its social and environmental consequences --- protection of the environment cannot be separated from the other objectives of energy policy" (Nathan, 1974).

ROLE OF ENVIRONMENT CANADA IN ENERGY

The role of Environment Canada in areas of energy development can be characterized by two major objectives.

- to influence the development and execution of policy and programs in energy in the public and private sector, to be in harmony with the environment; and

- to provide environmental knowledge and information to support the development and operation of safe, environmentally sound and economically efficient energy systems (Environment Canada, 1983).

The provision of environmental information obtained by collecting, analyzing and interpreting data at the national and international levels is an essential departmental contribution to federal energy policy and programs. Numerous departmental programs support the development and operation of alternative energy systems including the collection of information on weather, production of forest biomass as a source of energy, and the sensitivity of air, land, water and flora and fauna to energy developments. The Department has made a commitment to ensure that environmental information which is needed to support energy development is available and interpreted intelligently, and is applied in the design and operation of all energy systems. Environment Canada also discharges our responsibilities through regulation, intervention, persuasion and advocacy-methods which require a sound scientific and engineering understanding of the energy systems under consideration, and their environmental consequences combined with an appreciation of the economics and general socio-political considerations. The basic strategic approach in dealing with both the regulatory and advocacy options is the maintenance and expansion of a substantial body of relevant scientific and other knowledge and its application gleaned from appropriate forums such as this.

THE ROLE OF ENVIRONMENTAL PROTECTION SERVICE IN ENERGY

The Environmental Strategies Directorate of the Environmental Protection Service, in concert with a number of other government departments and agencies, has undertaken a series of fuel cycle assessments in order to provide a preliminary environmental assessment of existing and emerging energy technologies. Fuel cycle assessments undertaken to date include the thermal coal fuel cycle, frontier oil and gas, nuclear, and biomass fuel cycle. The need for a review and assessment of the biomass fuel cycle was precipitated by a number of developments which include: a recent shift in emphasis by the Department from a remedial to more of a preventative approach in dealing with environmental issues; research being undertaken by the government and private sector in energy areas in order to supplant our dependence on traditional non-renewable sources of energy; and the identification and legitimization of this assessment in national and international fora. A review of these national and international undertakings indicates that:

1) Little attention has been paid to a comprehensive analysis of the collection and processing techniques for biomass conversion to energy and the associated effects on the environment (Noyes Data Corporation, 1980).

2) A number of social, political, international, economic, technological, agricultural, and environmental constraints determine the limit of biomass energy development. No energy resource development should take place without a careful analysis of its benefits and cost to society (Pimental, D., 1983).

3) Large scale production of biomass energy can entail critical social, economical, and environmental considerations --- the development of biomass energy on a scale large enough to replace significant quantities of petroleum-based energy requires large capital investments and a commitment to a complicated support infrastructure. It is therefore important to both assess and consider the environmental consequences before large scale commitments are made (Marten, et al., 1981).

4) And finally, large scale utilization of biomass raises general questions of risk to the environment, including the preservation of soil nutrients and structure, erosion, water balance, methods of pest control, and the stability problems engendered by monoculture. Each of these questions must be answered well in advance of a commitment to significant utilization: a flow concept of energy production unlike the traditional reserve concept relies on the continued viability of the ecosystem from which the harvest is taken (Swain, et al., 1979).

BIOMASS FUEL CYCLE ASSESSMENT

The biomass fuel cycle assessment undertaken by the Environmental Protection Service is broken down into seven segments:

Part 1. Provides a development forecast of the standing amounts of biomass and other organic waste available for conversion from four diverse sources that include:

* silviculture (includes forest by-products, whole tree utilization and considers energy farms).

* aquaculture (considers the potential standing crops of macrocytic aquatic species as well as the growth of macrocytic and microalgae systems under four

different scenarios: open ocean systems, coastal systems, fresh water systems and integrated systems).

* agriculture (assesses conventional standing crops, crop residues, animal wastes, and agricultural energy systems).

* other organic wastes (municipal solid wastes, industrial wastes, sewage sludge, and food processing wastes).

Part 2. Discusses the environmental implications of the production and harvesting of these forms of biomass for energy. It considers:

* the environmental implications for soil, air quality, surface and ground water quality, wildlife habitat, and micro/macro climate.

* some mitigative measures that can be taken to address these environmental disruptions. For instance, in the bioconversion of forest products, the harvesting of forest biomass in winter to diminish the impact on soil and vegetation, or in summer to promote regeneration.

Part 3. Provides a description of some typical biomass raw materials and their conversion processes:

* examines the characteristics of typical biomass resources and provides a simple chemical breakdown and analysis.

* provides a generic overview of the three major conversion processes for the transformation of the biomass into energy:

* direct combustion

* thermo-chemical conversion -- pyrolysis, liquefaction, gasification

* biochemical conversion -- hydrolysis, anaerobic digestion fermentation

Part 4. Examines the environmental implications of the biomass conversion processes.

* Considers the potential for environmental disruption to air and water quality and solid waste emissions accruing from a number of different biomass conversion scenarios including:

* direct combustion of wood
* pyrolysis of wood
* incineration of municipal solid wastes
* anaerobic digestion of aquatic biomass

* Identifies some mitigative measures that exist in addressing those environmental concerns unique to biomass energy conversions. For instance, in the incineration of MSW, the pre-processing of wastes to provide:

* a reduction in volumes of refuse and more predictable, homogenous refuse

* a reduction in scavenger populations at land fill sites and partial elimination of obnoxious odors

* easier conveyor movement, magnetic separation, and air classification operations

Part 5. Provides an environmental overview of the production, transportation, handling and end-use bioenergy fuels:

* considers the air, water, solid waste and occupational health and safety concerns associated with large scale ethanol and methanol production facilities.

* some mitigative options that exist to address these potential environmental disruptions accruing at production facilities. For instance the recycling and reuse of sludge from wastewater treatment systems and ash from solid fuel combustion and processing for sale as animal feed - in a biomass to alcohol fuel facility.

* a generic environmental overview of transportation, handling and end use of biofuels.

* a preliminary assessment of the combustion of biofuels.

Part 6. Summarizes the Environmental issues associated with biomass/energy in Canada which:

* identifies those environmental issues unique to biomass energy scenarios.

* delineates those environmental impacts for which the knowledge of effects and control technologies is inadequate and which could be considered as major constraints to future biomass development.

* suggests direction for the associated research and development activities.

Part 7. Provides an overview of the legislative dimensions associated with biomass/energy development in Canada which:

* examines the Environment Canada mandate in relation to biomass to energy development.

* examines a list of non-legislative options available to Environment Canada to influence biomass development in an environmentally sound manner.

* provides a list of Provincial Acts, regulations, and procedures relevant to biomass to fuel development.

An overriding qualifier must be considered in use of this document:

i) This document does not constitute a departmental position on biomass nor is it the definitive statement on the environmental impacts anticipated from large scale utilization of biomass for energy. It emphasizes the need for the consideration of the environmental parameters early in and at each stage in the development process of any biomass/energy scenario well as identifies some preliminary departmental concerns in these areas. As more research into some of the environmental data gaps associated with various bioenergy scenarios becomes available, it will be integrated into revised and updated versions of this study.

In addition this paper does not consider those environmental concerns associated with biomass conversion to petro-chemical feedstocks.

OVERVIEW OF GENERIC ENVIRONMENTAL CONCERNS FROM LARGE SCALE BIOENERGY DEVELOPMENT

Time constraints limit me from identifying all areas of environmental concern as identified in this document. However, based on this preliminary assessment and a review of analogous undertakings; I would draw your attention to some outstanding generic environmental considerations that still require further analysis in any bioenergy undertakings. These include: (US Department of Energy, et al., 1977)

i) The large land and water requirements of terrestrial and aquatic aquatic biomass plantations may restrict the development of competing uses of land and water in some regions and cause adverse socio-economic impacts. (This is viewed as a possible concern in the long term and not an acute issue). While agricultural and silvicultural needs have been addressed, research on resource requirements for biomass plantations is lacking.

More specifically, a region by region study of land and water availability, biomass land and water requirements, and assessment of impacts of competition between biomass production and other uses of land and water is needed.

ii) Biomass plantations (aquatic, agricultural and silvicultural) have the potential to alter significantly air and water quality in a number of ways. Research on air and water pollution from agricultural and silvicultural sources should be extended to the pollution implications from aquatic plantations. Research already conducted for agriculture and silviculture should be reviewed in an environmental context.

iii) Removal of biomass from aquatic and terrestrial ecosystems may pose environmental implications, which may limit future crop growth. The extent to which these environmental problems may occur, or the amount of biomass that might be removed before damage is caused has not been determined for most of the materials considered as having biofuel potential.

iv) Direct combustion of biomass can, in some instances, emit greater quantities of the air pollutants like nitrogen dioxide, carbon monoxide, and particulates than fossil fuel combustion. Information and research on characterization of and control strategies for these pollutants is necessary. Such research should include: a review of existing research on air pollution and associated control methods for wood boilers; a study to develop pollution control technology for biomass combustion plants; and a study to monitor air pollutants emitted from representative direct combustion biomass plants.

v) The various thermochemical biomass conversion processes produce small quantities of pollutants such as hydrogen sulfide and phenols which can affect air and water quality. In addition, tars and oils generated from the pyrolysis process may present considerable health and safety concerns. Attention should be directed towards specific waste streams from thermochemical processes in terms of pollutant monitoring; definition and characterization; development of controls; and waste disposal methods. Similar attention should also be directed to biological conversion processes as well.

vi) The anerobic digestion processes produce a sludge which must be disposed of in an environmentally sound fashion to avoid pollution of surface and groundwater. Particular attention should be directed towards characterization of the residue and possible alternative uses for the sludge.

I would also draw your attention to the broad array of generic socio-economic impacts which have to be reconciled in any major shift to a biomass fuel economy. These would include in part:

GENERIC SOCIO-ECONOMIC IMPACTS OF BIOMASS ENERGY SYSTEMS

Area of Interest	Impact or Issue Description
Land Use	Land requirements for biomass energy production are orders of magnitude larger than for other energy technologies. This raises significant aesthetic and practical impact and land-availability questions. Severe conflicts among competing land-use interests can be expected.
Market linkage	Linkage among biomass product markets can be complex and can lead to either increasing or decreasing overall market stability. Diversity in feedstock and product mix and vertical or horizontal process integration should probably be encouraged.
Dynamics of market expansion	Excessive subsidization, stored resources and probable expansion of more-profitable biomass markets could imply wasteful and destructive overexpansion of biomass energy production.
Food	Competition between food and energy biomass could affect domestic and international food supplies and prices.
Employment and demographic changes	Location of biomass facilities in rural ares would generally produce significant employment and community impacts. Regional demographic shifts may also occur as a result of variations in biomass availability.
Institutional issues	Relationships among biomass energy producers, existing utilities, and potential customers for cogenerated steam would need to be developed. Environmental regulations, tax treatment and government revenue implications, and capital availability are also major areas of uncertainty.
Political issues	Decentralization and subsidies among competing energy programs may become even more prominent political issues.
Transportation	Road upgrading and maintenance requirements as well as traffic pattern changes may be significant. Indirect impacts of transportation system alterations should also be considered.

Source: Braunstein, H., et al., (1981) "Biomass Systems and the Environment", Oakridge National Laboratory, Pergammon Press, New York, p. 153.

SUMMARY

To recapitulate, I would ask you to consider the following factors:

1. That biomass/energy conversion schemes are complex and diverse undertakings that require in-concert consideration of those parameters of technological soundness, economical viability, environmentally benignness, and social acceptability.

2. That the potential for environmental disruption associated with each form of biomass considered for conversion to energy should be addressed at each stage in the fuel cycle. The diversity of biomass fuel stocks, coupled with the multiplicity and flexibility of conversion processes will probably require more of a site-specific environmental assessment.

3. That biomass energy is only renewable within specified limits. Mismanagement at any of a number of stages in the energy conversion process could exhaust many of the biomass supplies considered. Our soil and water resources are limiting factors in the amounts of energy sources that might safely and perpetually be harvested. Exceeding these limits will deplete the resource base and diminish any available biomass supply (Pimental, et al., 1981).

CONCLUSION

In conclusion, I would like to say that our present understanding of energy related environmental hazards is sufficient to permit far better integration of environmental criteria into the engineering, design, and operation of new energy options. Direct harnessing of sunlight, energy from biomass and increased efficiency of energy represent a diverse array of technological possibilities, and it is essential to emphasize the development of the most environmentally enhancing options in the development of energy from biomass.

But this opportunity only exists if we abandon the archaic view of solving energy problems which has historically pervaded energy programs. That is, as a simple matter of expanding energy supply irrespective of environmental and socio-economic considerations.

REFERENCES

1. Lord Nathan (1974) "Energy and the Environment", pulished on behalf of the Committee for Environmental Conservation, the Royal Society of Arts, and the Institute of Fuel, London, England.

2. Corporate Planning Group, Environment Canada (1983) "Energy and the Environment - Environment Canada's Energy Strategy" Annex 2, pp. 9.

3. Noyes Data Corporation, (1980) "Fuels from Biomass - Technology and Feasibility" ed. by J.G. Robinson, Park Ridge, New Jersey. Energy Technology Review 6I, Chemical Technology Review 176, 80-23488, pp. 377.

4. Pimental, D. (1983) "Review Article - Biomass Energy", the Biomass Panel of the Energy Research Advisory Board, in <u>Solar Energy</u>, Vol. 30, No. 1, pp. 1-31, 1983, Great Britain.

5. Marten, G., et al., (1981) "Environmental Considerations for Biomass Energy Development: Hawaii Case Study", East-West Environment and Policy Institute, Research Report No. 9, East-West Center, Honolulu, Hawaii, pp. 59.

6. Swain, H., Overend, R. and Ledwell, T.S. (1979) "Review Paper - Canadian Renewable Energy Prospects", in Solar Energy, V. 23, pp. 459-470, Pergammon Press, Ltd. 1979, Great Britain.

7. United States Department of Energy (1977) "Environmental Development Plans - Fuels from Biomass" US-DOE/EDP-0005, US 11. 41, 61, Washington, D.C. See also Hruby, T. (1978), "Impacts of Large Scale Aquatic Biomass Systems" Woods Hole Oceanographic Institution, Woods Hole Massachussetts, 02543 (March 1978), Technology Report, and Braunstein, H., et al., (1981) "Biomass Systems and the Environment",

8. See Pimental D. "Biomass Energy - Bonanza or Boondoggle" in Soft Energy Notes, Dec/Jan 1981, pp. 8-10. For overviews from a number of perspectives see also: Bungay, H. "Commercializing Biomass Conversions" in Env. Sci. & Tech., Vol 17, N. 1, 1983, pp. 24A-31A; Rom, W.N, and Lee J. "Energy Alternatives: What Are Their Possible Health Effects?" in Env. Sci. and Tech., Vol. 17, N. 3, 1983, pp. 132A-143A; Hira A., et al., "Alcohol Fuels from Biomass - An Overview", in Env. Sci. & Tech., Vol. 17, N. 5, 1983, pp. 202A-213A. Oakridge National Laboratory, Pergammon Press, New York.

CONTINUOUS ETHANOL PRODUCTION: FERMENTATION AND PURIFICATION IN A SINGLE VESSEL

D. Lindsay Mulholland

Ontario Research Foundation
Sheridan Park Research Community
Mississauga, Ontario, Canada

ABSTRACT

The cost effectiveness of continuous systems compared to repeated batch operations has led scientists and engineers to scrutinize the traditional fermentation/distillation process for alternatives which will eventually lead to simpler and more economic production. The work presented here covers the development, by Ontario Research Foundation, of an embryonic technology which could potentially meet the goals of reduced alcohol plant capital and operating costs.

The proposed technology is suitable for systems where the product of fermentation is sufficiently volatile that fermentation and recovery from the broth can take place simultaneously. A recycled stripping gas is used for this purpose. The apparatus is a cylindrical vessel fitted with an open-ended cylindrical draft tube which sets up a cyclical flow pattern in the liquid. For ethanol fermentation, the gas becomes saturated with alcohol and water vapour, passes through a condenser and is reintroduced to the column via a porous inlet.

Initial work on the establishment of equilibrium and kinetic data on the system, simply as an ethanol purification device, is reported. This is followed by a discussion on the use of the system as a viable fermenter with Saccharomyces spp. In its role as a fermenter, the apparatus has operated continuously for four months. Yields and rates of production of ethanol are reported.

INTRODUCTION

Presently, marginal economics and low profitability are delaying the development of a fuel alcohol industry in the industrialised world. A great deal of attention, by the engineering community, is being focused on the purification stage where dilute alcohol beers are concentrated to anhydrous fuels. The energy requirement to produce anhydrous ethanol from a 10% by weight beer is about 25% of the energy content of the ethanol produced (Black, 1980). Distillers of potable alcohol are also looking for ways to reduce the energy expenditure involved in fractional distillation (Safty, 1984). Several techniques are being explored as possible replacements for the traditional energy-intensive distillation of dilute alcohol beers. These include: low pressure distillation, vapour recompression, extractive distillation, liquid-liquid or solvent extraction, adsorption (molecular sieves), dehydration using absorbent solids, and reverse osmosis. Several reviews of these methods have recently been published (Black, 1980; Field, 1981; Morris, 1981). This paper describes another promising technique as a possible replacement for energy-intensive distillation.

This technique is based on gas stripping of the fermentation beer in a specially-designed column. This is illustrated schematically in Figure 1.

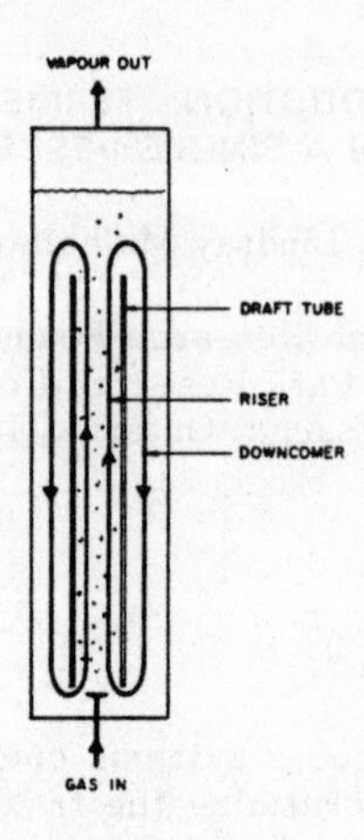

Figure 1 Typical configuration of gas liquid contacting device with draft tube.

To our knowledge, no information is available in the literature, regarding either the vapour-liquid equilibrium or the kinetics of gas stripping of aqueous ethanol solutions, at temperatures lower than boiling. In order to investigate the commercial feasibility of this technique, it was therefore necessary to establish the pertinent data background.

EXPERIMENTAL

The apparatus used was constructed with the dimensions shown in Figure 2. The column walls were made of quartz glass and the inner draft tube of stainless steel. The draft tube was made double-walled to allow passage of heat transfer fluids for regulating the temperature of the column contents. Separate stainless tubes were attached and passed through the base plate of the apparatus for this purpose. This arrangement was later used to carry steam for sterilization of the column for the fermentation studies.

The exiting gases from the column were passed into a series of three high-efficiency laboratory glass condensers which were cooled to 6± 1°C. Uncondensed ethanol and water were continuously assayed by a gas phase infrared spectrometer (Miran analyser) and, during physico-chemical testing, were then vented. During fermentation, the infrared gas analyser was omitted and the gases from the condensers were returned to the column by a gas pump in a closed cycle. The condensate was collected in a 2L round-bottomed laboratory flask fitted with a side arm attachment to allow passage of the uncondensed gases. The ethanol content of the condensate was analysed by gas chromatography.

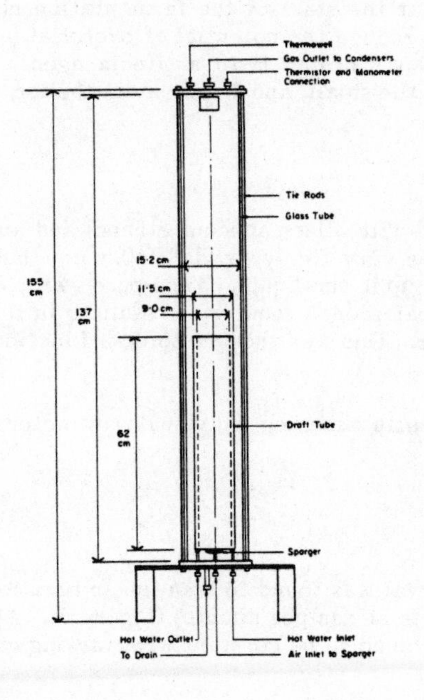

Figure 2 Experimental apparatus for low cost recovery of ethanol.

Fermentation was carried out using two strains of Saccharomyces cerevisiae ATCC 4126 and 4132. Initial experimentation in continuous culture took place in 500 mL Erlenmeyer flasks fitted with sintered glass sparging inlets. Nutrient solutions were added by sterile syringe through a hypodermic needle permanently fitted to the rubber stopper which acted as the vessel closure. Broth samples were also removed by the syringe needle. The stopper was fitted with a glass tube to allow passage of the exiting gases. The gases were passed through a single Liebig condenser. The condensate was collected in a 100 mL round-bottomed flask and the uncondensed gases were vented. The Erlenmeyer flasks were immersed in water in a thermostated bath. The large-scale continuous fermentations were carried out in the draft tube column described above. The entire volume of the column was approximately 20 L, but an effective working volume of 11 L was used. This was sufficient to cover the draft tube and afforded a sizeable headspace should foaming present a problem. The headspace was externally heated to 40°C to prevent condensation of the volatiles. Organism culture and nutrition were by normal microbiological techniques. The carbon source was glucose and the other mineral and trace requirements were provided by the chemicals KH_2PO_4, K_2HPO_4, citric acid, sodium citrate, $MgSO_4$, $(NH_4)_2SO_4$ and yeast extract.

The growth of the culture was monitored by optical density at the early stages and by ATP assay during the extended continuous runs. This is the luciferin-luciferase (firefly) enzyme assay, where the bioluminescent response is a linear function of the ATP concentration of a quantitative extraction of the cells.

The sparging gas was bottled CO_2 for the small-scale experiments replacing bottled air which was used in the exponential growth phase of the runs. Fermentation gas was recycled with a diaphragm pump delivering approximately 18 Lpm in the large-scale column fermenter. The pH was 5.5 at the start of the fermentation but was allowed to decrease during the continuous runs to reduce the potential of bacterial infection. Foaming was prevented by the addition of Dow-Corning H-10 antifoam agent. The rate of addition was approximately 150 µL/day in the small, and 1 mL/day in the large vessel.

RESULTS AND DISCUSSION

The apparatus was tested with dilute aqueous ethanol and air as the stripping gas. The gas bubbles were found to be very finely divided (<0.5 mm bubble diameter) and a good cyclical flow pattern was set up in the liquid. When pure water or very high concentrations of ethanol were employed, coalescence took place resulting in the formation of gas slugs of approximately 1 cm diameter. This was thought to be a function of increased surface tension.

The effects of independently changing individual parameters are discussed under separate headings below

Effects of Gas Flowrate

The rate of ethanol removal was found to be a linear function of gas flowrate from 0.1 to 2.0 vvm (volume per volume of gas per minute) (Figure 3). Also, the ethanol concentration in the vapour phase was found to be constant with varying gas flow (Figure 4).

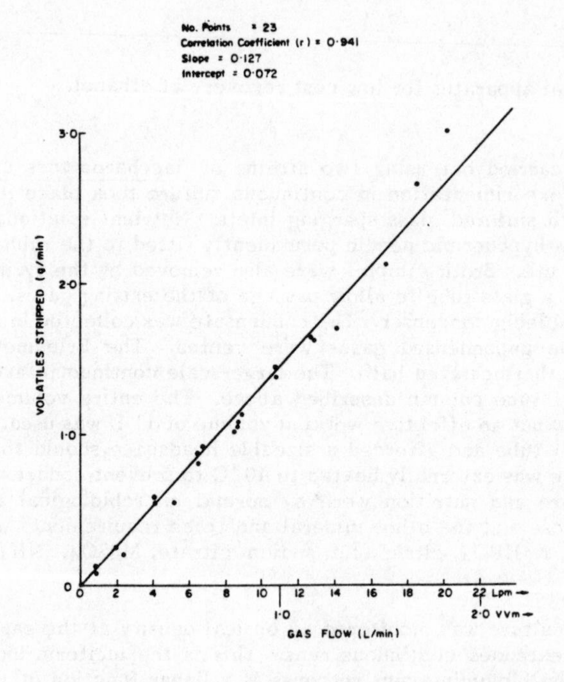

Figure 3 Total volatiles stripped vs. gas flow

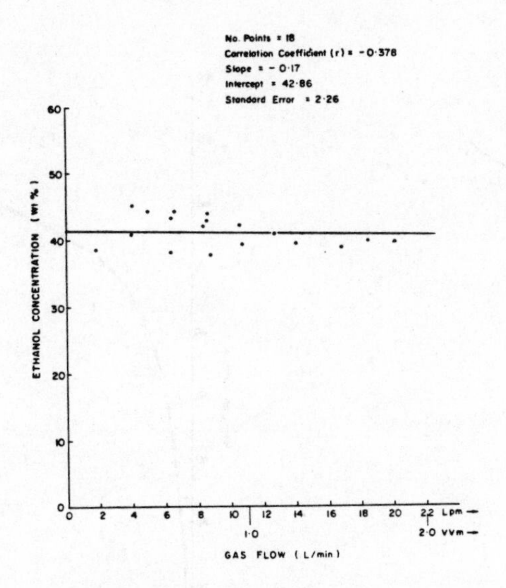

Figure 4 Ethanol concentration in the vapour phase vs gas flow.

Effects of Nature of Sparging Gas

No changes were observed in the vapour composition of volatiles or in condensation rates when the sparging gas was changed. Gases tested were air, helium, carbon dioxide and nitrogen. This suprising result is a reflection of near ideal behaviour in the gas phase of the volatile components under the operating conditions of low temperature and pressure (see below).

Effects of Ethanol Concentration in the Liquid Phase

The vapour-liquid curve generated at 40°C in this work was closely similar to that obtained at the boiling point and reported in the literature; Figure 6 compares well with Figure 5. It was, therefore, concluded that the ratio of ethanol to water in the vapour phase depended only on the concentration in the liquid phase over the temperature range of observation (40°-100°C). The influence of ethanol concentration on the rate of evaporation of the volatiles will be dealt with in a later section.

The Rate of Evaporation and Condensation

A mathematical relation has been derived which describes the rate of volatile uptake with varying concentration. This is based on a modification of the absolute humidity expression which is derived from the gas laws, the only assumption being that ideality is

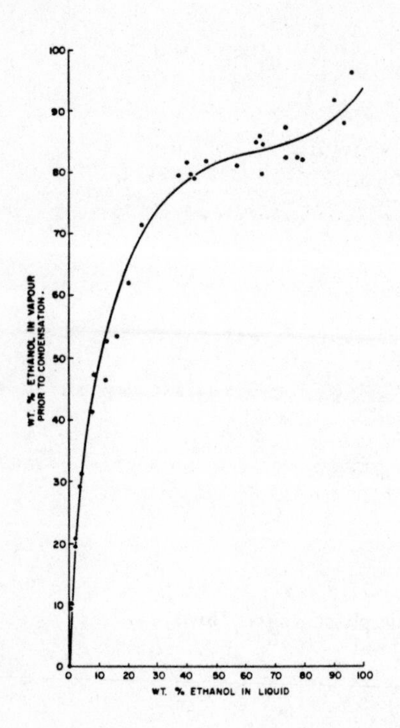

Figure 5 Vapour-liquid curve (this work).

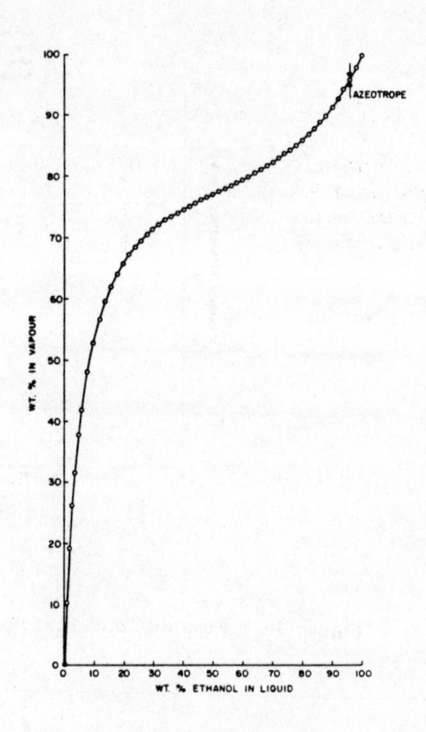

Figure 6 Vapour-liquid equilibrium curve from literature.

approximated in the vapour phase. The derived expression is

$$\sigma = \frac{1.4\ M_{av}\Sigma fi}{M_g\ (P - \Sigma fi)}$$

where σ is the specific volatile uptake; 1.4 is a constant for the gas used, here, it is the density of moist air;

M_{av} is the "averaged" molecular weight, i.e., $M_{av} = \Sigma x_i MW_i$, where x_i and MW_i are the mole fraction and molecular weight of the component i of the mixture respectively.

Σfi is the sum of the fugacities of the components where fi, the fugacity of the i-th component, is calculated from $fi = x_i \sigma_i P_i{}^\circ$, where x_i, is the mole fraction of the i-th component in the liquid, σ_i is the activity coefficient of the i-th component obtained independently, $P_i{}^\circ$ is the vapour pressure of the pure component i at the same conditions of temperature and pressure.

M_g is the molecular weight of the stripping gas, and P is the total system pressure (usually 1 atmosphere).

The mass transfer rates from the liquid to the vapor phase are obtained from the relation:

$$r = f\sigma$$

where r is the volatilization rate, and f is the gas flowrate.

Excellent agreement was found between calculated and experimental results up to ethanol concentrations of about 80 wt% (Figure 7). The apparent deviation at higher ethanol concentrations may be attributed to poor activity coefficient data for water at very low concentrations.

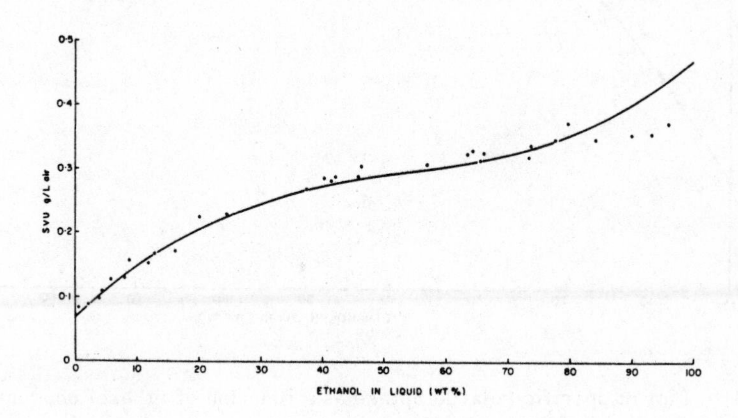

Figure 7 Plot of specific volatile uptake as a function of ethanol concentration at 40°C. The line is from calculation, the points are experimental.

The Rate as a Function of Temperature

Experimentally, it was found that the specific volatile uptake approximately doubled for each 10°C rise in temperature in the range 40-60°C. Results calculated using the vapour pressures of pure ethanol and water at 50° and 60°C instead of 40°C gave values in reasonable agreement with the experimental results (Figures 8 and 9).

Fermentation Studies

Preliminary work was carried out with small-scale laboratory glassware. Two yeast strains ATCC 4126 and 4132 were cultured and grown with and without vigorous gas sparging. Volatiles were returned to the broth by a reflux condenser. The results are illustrated graphically in Figure 10. The experiments were conducted at 25 and 40°C. Growth was monitored by turbidimetry and ethanol and glucose levels were assayed by gas chromatography and High Performance Liquid Chromatography (HPLC) respectively. Results are shown in Figures 11 and 12.

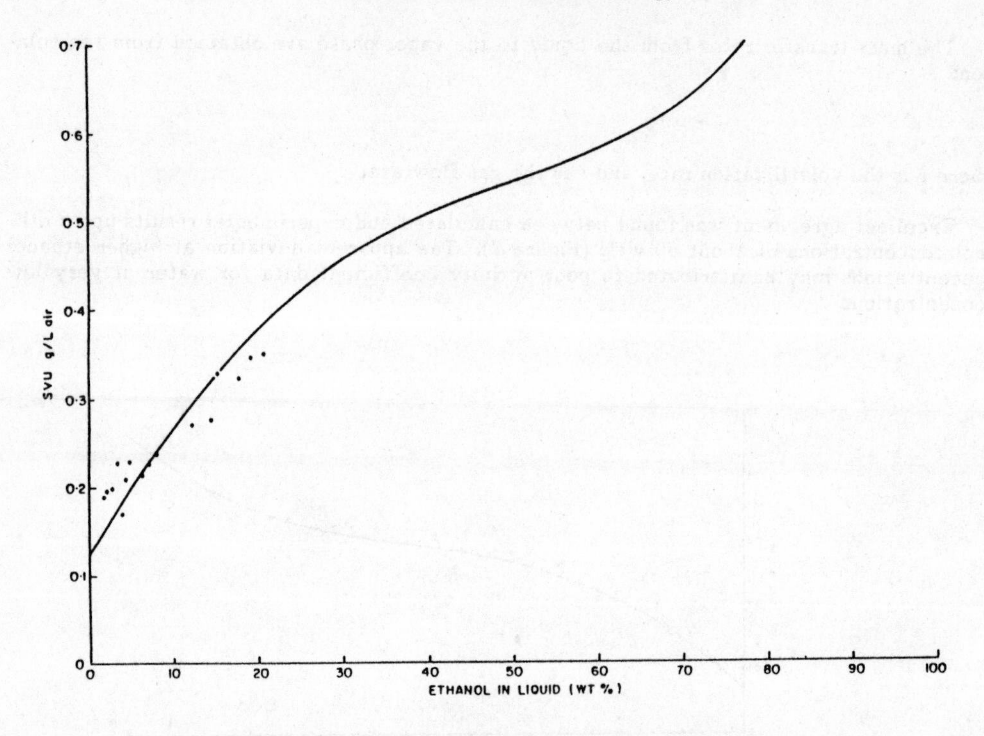

Figure 8 Plot of specific volatile uptake as a function of ethanol concentration at 50°C. The line is from calculation, the points are experimental.

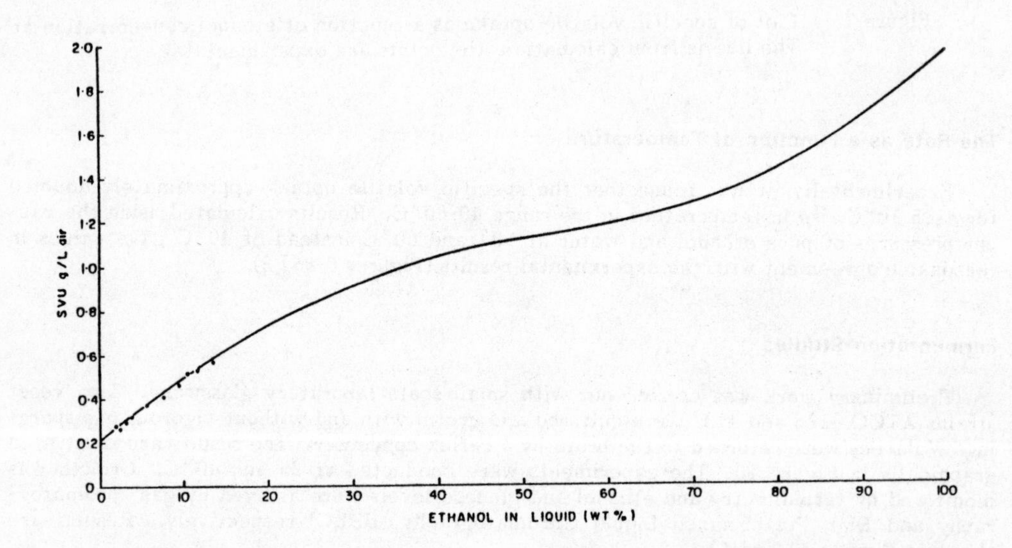

Figure 9 Plot of specific volatile uptake as a function of ethanol concentration at 60°C. The line is from calculation, the points are experimental.

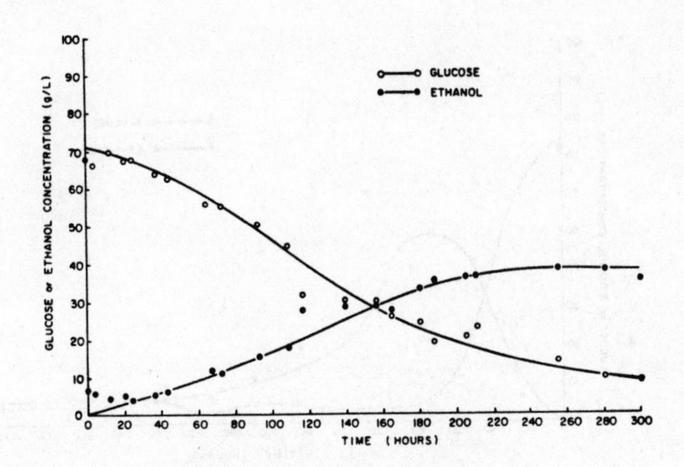

Figure 10a Glucose and ethanol concentrations with time-quiescent run.

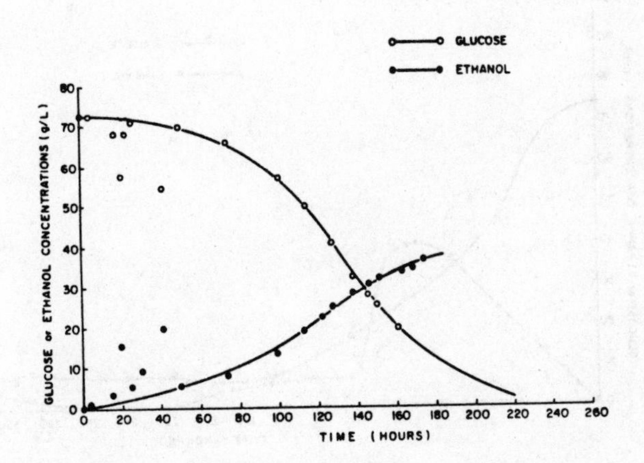

Figure 10B Glucose and ethanol concentrations with time-sparged run.

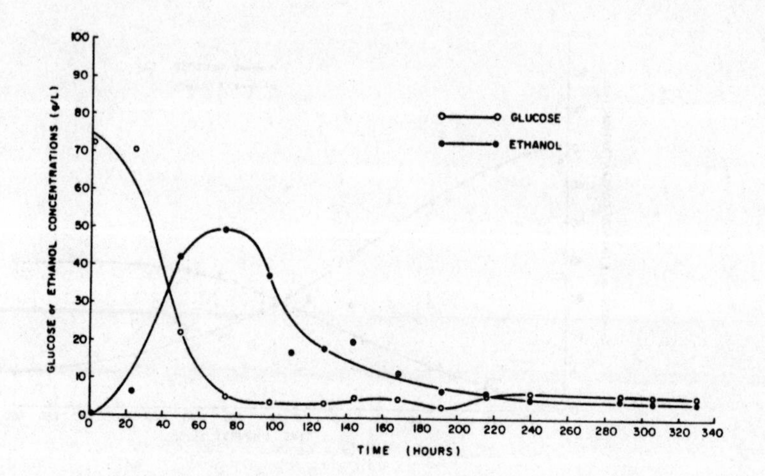

Figure 11A Fermentation with strain 4132 at 25°C with continuous gas stripping.

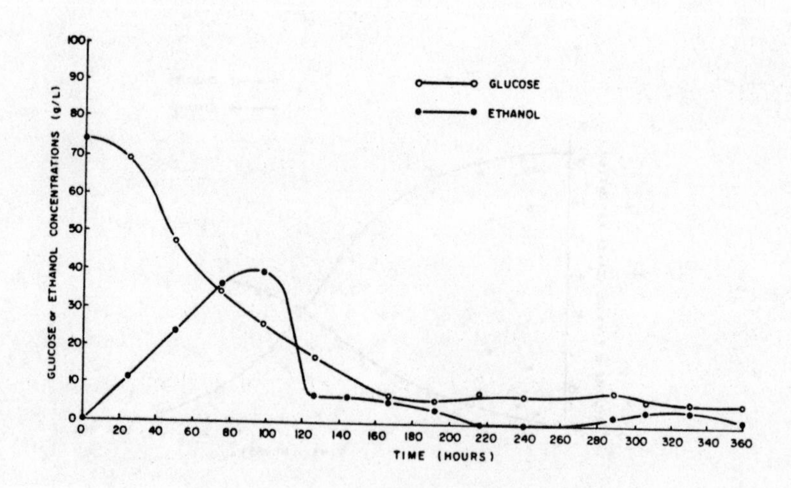

Figure 11B Fermentation with strain 4132 at 40°C with continuous gas stripping.

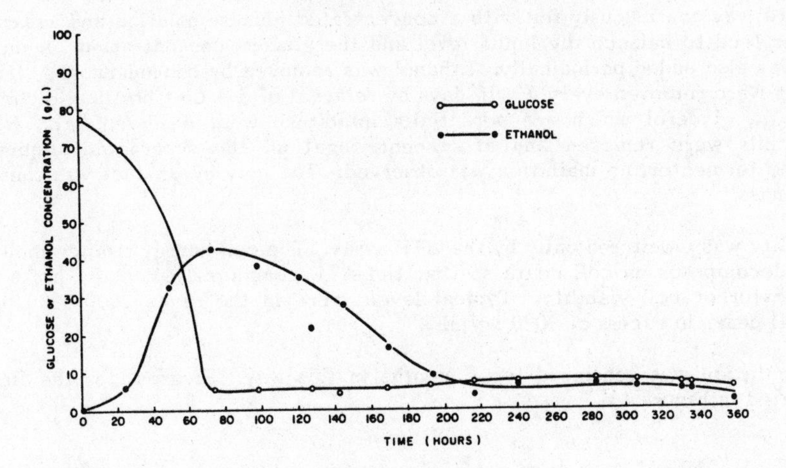

Figure 12A Fermentation with strain 4126 at 25°C with continuous gas stripping.

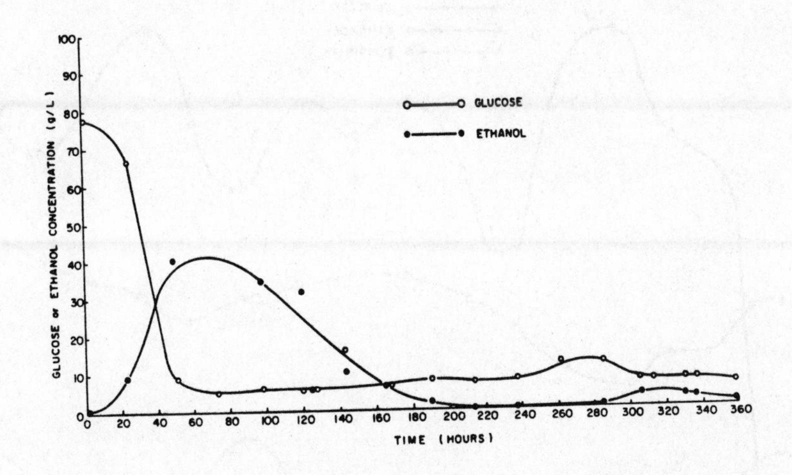

Figure 12B Fermentation with strain 4126 at 40°C with continuous gas stripping.

These experiments demonstrated that

(a) growth and fermentability of both strains were not inhibited by vigorous gas sparging of either air, nitrogen, or CO_2,

(b) the growth and fermentability of strain 4126 was very similar at 40°C and 25°C under conditions of vigorous gas sparging, whereas

(c) the growth and fermentability of strain 4132 was severely inhibited at 40°C compared to identical experiments at 25°C.

These results led to the selection of Saccharomyces cerevisiae strain ATCC 4126 as our test organism. Further work was carried out in the 11 L apparatus illustrated in Figure 2.

The culture was continuously fed with a concentrated glucose solution and separately, a sterile water feed to balance the liquid level and the glucose concentration. A nutrient concentrate was also added periodically. Ethanol was removed by continuous gas stripping and spent cells were removed every 7 - 10 days by removal of 3-4 L of broth. No chemical by-products e.g., glycerol which are potentially inhibitory were observed (g.c., HPLC). When spent cells were removed and after centrifugation, the supernatant liquid was returned to the fermenter, no inhibition was observed. The only by-product was therefore spent cell biomass.

Cell viability was monitored daily by the ATP assay. The cell energy storage molecule, ATP, rapidly decomposes on cell death so that the ATP concentration in the broth is an excellent indicator of cell viability. Typical levels were in the range 3000-5000 ng/mL with occasional peaks in excess of 9000 ng/mL.

The fermentation was continued for 4 months in this way. Progress for the first 16 days is illustrated in Figure 13.

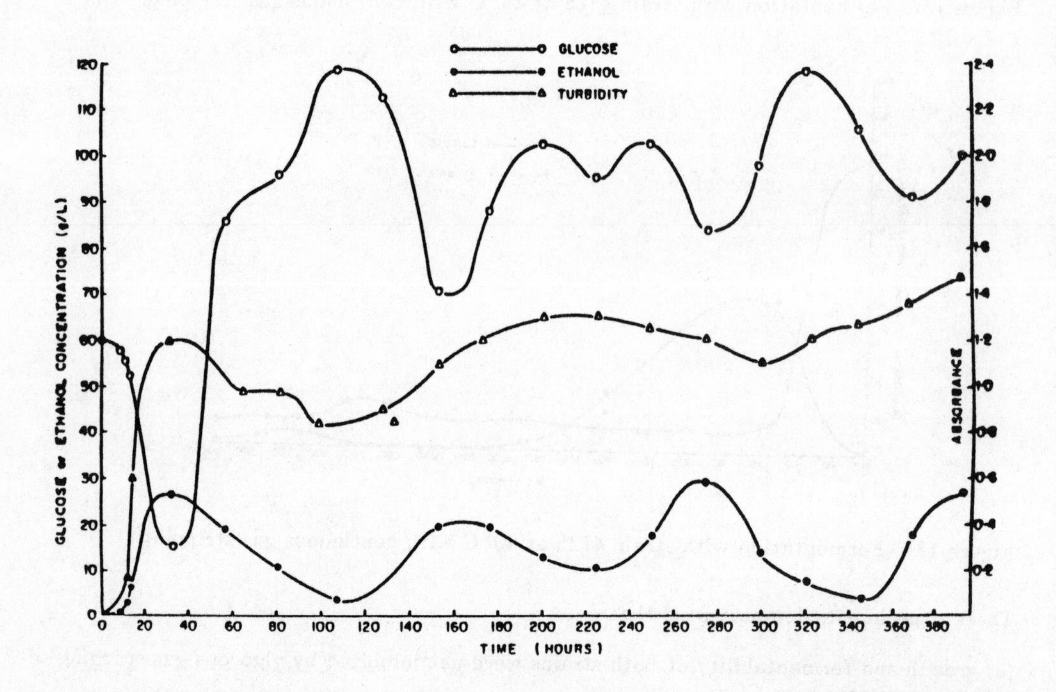

Figure 13 Fermentation with strain 4126 in column fermenter at 40°C with continuous gas stripping.

The yield and rate figures for the first one-month segment of the fermentation were calculated from the data listed below:

• Duration of fermentation run (this segment)	789 h
• Glucose consumed (total in experiment)	9,050 g
• Volume condensate collected	55.3 L
• Concentration of pooled condensate	77.3 g/L
• Fermenter volume	11 L

Yield

• Ethanol produced (55.3 x 77.3)	= 4,274.69 g
• Theoretical yield (.511 x 9,050 g)	= 4,624.55 g
• Percentage yield	= 92.43%

Rate

• Ethanol production rate (4,274.69/789 h)	= 5.418 g/h
• Ethanol productivity (5.418/11)	= 0.49 g/L-h
• Hourly specific dilution rate (55.3/[11 x 789 h])	= 0.006 h^{-1}

Biomass

During this period, the biomass concentration varied between 7 and 27 g/L on a dry weight basis. These were determined by removing two 20 mL samples, centrifuging, removing the supernatant, transferring the biomass sludge to a preweighed aluminum dish with washings sufficient to complete the transfer, and evaporating the water in an oven at 105 °C until constant weight. The mean of duplicate determinations was used as the biomass concentration in units of weight of oven-dried biomass per unit volume.

From these and other measured values, it was estimated that the biomass production rate was 0.26 g/h for the 11 L fermenter, i.e. the specific biomass production rate was 0.023 g/L-h. This figure is approximately 5% of the ethanol production rate (0.49 g/L-h), i.e. for each metric tonne of ethanol produced, there will be 46.7 kg of biomass.

Alternatively, each tonne of glucose will produce:

472.34 kg ethanol
451.98 kg CO_2 (assuming ethanol:CO_2 is 1:1)
22.06 kg biomass

The ethanol yield is high. During the period of fermentation, four dilutions of the broth were carried out, resulting in a loss of feedstock, product and biomass. While these were not large losses in relation to the total activity of the fermentation, they did contribute to the lowering of the yield figure from closer to 100%. Some glucose for cell building is inevitably necessary, but cell turnover in a continuous culture should ideally be low.

The rate of ethanol production and the hourly dilution rates are very low. The specific hourly production rate is approximately 1/4 to 1/3 that of a batch fermentation. In practice, it is probably comparable if down-time (clean-out time) is calculated for the batch mode. Spoiled batches also often contribute to a lowering of the 1.5-4 g/L-h figure typical of batch fermentations. The hourly dilution rate is extremely low, indicating a low cell productivity.

CONCLUSIONS

On a laboratory scale, gas stripping of dilute ethanol solutions has been explored. An expression has been derived by which prediction of volatile uptakes and condensation efficiencies can be calculated over the temperature range of interest (40-60°C). The technique holds considerable promise for the continuous stripping of continuous fermentations.

The feasibility of long term continuous fermentation has been demonstrated with a common yeast which was shown to grow and ferment under high temperature conditions (40°C) when the broth was subjected to gas sparging. Other organisms may prove more efficient under these conditions, and the possibility exists for using a thermotolerant organism which would allow an even higher temperature operation.

REFERENCES

1. Black, C., Distillation Modelling of Ethanol Recovery and Dehydration Processes for Ethanol and Gasohol. Chem. Eng. Prog., September, 1980, 78-85.

2. Field, E.L. U.S. Patent No. 4,303,478 (1981).

3. Morris, E.C., Food Eng., Feb., 1981, p. 93.

4. Hone, R.W., Lamarchand, M., and Malaty, W., Separation of Water-Ethanol Mixtures by Sorption, Part 2, Oak Ridge National Laboratory, Feb. 1981.

5. (a) Scheibel, E.G., Ind. Eng. Chem. 42; 1497(1950).
(b) Roddy, J.W., Ind. Eng. Chem. Proc. Res. Dev. 20, 104 (1981).
(c) Leeper, S.A., Wankat, P.C., ibid 21, 331 (1982).
(d) Roth, E.R., US Patent No. 4,306,884, December 1981.

6. Eakin, D.E., et al., Preliminary Evaluation of Alternative Ethanol/Water Separation Processes, Batelle Memorial Institute, May, 1981.

7. Sheppard, J.D., and Margaritis, Biotech Bioeng. 23, 2117 (1981).

8. Cornell, L.W., and Montonna, R.E., Ind. Eng. Chem. 25, 1331 (1933).

9. Safty, M. McGuinness Distillers. Personal communication (1984).

ANAEROBIC CONVERSION OF PRETREATED LIGNOCELLULOSIC RESIDUES TO ACIDS

Silvia Ortiz

Applied Research Division - ICAITI
Central American Research Institute for Industry
Guatemala City, Guatemala

Utilization of agroindustrial residues and wastes with high levels of lignin, hemicellulose and cellulose from tropical regions, particularly Central America, is an area of great interest to ICAITI.

We are searching for alternative uses of the lignocellulosic residues, first separating the wastes into their principal macro-components through a series of pretreatments and then using the resulting materials for the production of ethanol, biomass or fuel.

Bagasse from citronella (Cymbopogon nardus), lemon tea (Cymbopogon citratus or C. flexuosus) and sugar cane (Saccharum officinarum) are the substrates in this study. Their contents of lignin, cellulose, hemicellulose and protein are presented in Table 1. Lemon tea bagasse as well as citronella bagasse are wastes from the distillation of essential oils. Samples collected in the distilleries were sun dried to 6-7% moisture and stored in plastic bags. Sugar cane was collected in a sugar mill and processed in ICAITI by the EX-FERM process (Rolz et al., 1979). The fermented chips were washed with water, pressed and sun dried to 6-7% moisture.

Table 1. Composition of the Lignocellulosic Substrate Used

	Dry Base			
	Lignin	Cellulose	Hemicellulose	Protein
Sugar cane bagasse	12.5	48.5	17.0	1.2
Citronella tea bagasse	11.1	30.2	15.5	4.5
Lemon tea bagasse	11.0	30.0	5.1	9.8

Different pretreatments for these lignocellulosic residues were evaluated. The pretreatment methods are:

1. Methods With High Pressure and Temperature

1.1 Conventional soda cook:

- Cooking temperature: 160°C
- Time needed to reach cooking temperature: 90 minutes
- Cooking time: 90 minutes
- Liquid to solid ratio: 1:5
- Sodium hydroxide: 15% of raw material on a dry weight basis

1.2 Ethanol-water mixture used in the presence of sodium hydroxide and anthraquinone:

- Cooking temperature: 175°C
- Cooking cycle: 240 minutes
- Liquid to solid ratio: 1:6
- Ethanol by weight: 60%
- Water by weight: 40%

1.2.1 The first run was carried out by adding 1% sodium hydroxide to the cooking liquor (1% of the dry weight of the raw material). After cooking, each unit was quickly cooled submerging it in water. The cooked material was fiberized in a two-liter blender, together with the cooking liquor. The mass thus obtained was separated from the cooking liquor and air dried. The dry material was ground in a Fitz-Patrick mill to pass a 1 mm screen.

1.2.2 The second run was similar except that the quantity of sodium hydroxide was increased to two percent and that the cooked material was separated from the cooking liquor and fiberized using fresh water. As in the previous run, the material was air dried and ground.

1.2.3 The third run was the same as the second run except that 0.05% anthraquinone was added to the cooking liquor as a catalyst.

1.2.4 In the fourth run, sodium hydroxide was four percent and 0.05% anthraquinone was added to the cooking liquor.

2. Methods Using Gases:

2.1 Sulfur dioxide:

- Whole sugar cane bagasse, lemon grass bagasse and citronella grass bagasse were ground to pass a 1 mm screen
- Particles were moistened with distilled water using a solid to liquid ratio of 1:3
- Materials placed in Buchner funnels were exposed to a stream of SO_2 until saturated
- Material was placed in a sealed jar and incubated at 70°C for 72 hours - Products were air dried and transferred to polyethylene bags.

2.2 Gaseous ammonia

- Raw materials were ground to pass a 1 mm screen and moistened with distilled water using a solid to liquid ratio of 1:3
- Materials placed in Buchner funnels were saturated by passing gaseous NH_3

- Material was transferred to a polyethylene bag and maintained for ten days at room temperature
- Products were air dried and stored in polyethylene bags

3. Methods Using Alkali:

3.1 Sodium hydroxide:

- Eighty grams of sodium hydroxide was added to each kg of raw material (dry basis)
- Solid to liquid ratio was 1:4
- Well mixed ground material and the sodium hydroxide solution was placed in a polyethylene bag for 10 days at room temperature
- Product was air dried

3.2 Sodium carbonate and calcium hydroxide mixture:

- Ground material was mixed with sodium carbonate at 0.663%
- Calcium hydroxide was slurried in water and added at a level of 0.25% of raw material
- Sufficient water was added to give a solid to liquid ratio of 1:4
- After thorough mixing, the material was left at room temperature for ten days in a polyethylene bag
- Product was air dried

4. Biological Methods:

- The raw material was moistened to contain 25% water and innoculated with Basidiomycetes
- Incubation at room temperature was 4-6 weeks, with aeration every 4 days
- Product was air dried

In vitro digestibility studies evaluated the effects of these pretreatments on the biodegradability of substrates (2,3).

After the lignocellulosic material is treated, conversion to organic acids is carried out using one of two procedures. The first is conversion of the holocellulose to acids using mixtures of cellulolytic and non-cellulolytic bacteria obtained from different culture collections. At this time dry weight of the remaining substrate was checked to determine utilization by the bacteria, as well as acidity, volatile acid production, pH and total gas production.

Table 2. Pure Cultures Used in this Study

Acetovibrio cellulolyticus ATCC 33288
Ruminococcus albus ATCC 27210, NCDO 2250
Clostridium thermocellum ATCC 27405
Clostridium thermosaccharolyticum ATCC 7956
Ruminococcus flavefaciens ATCC 2213
Bacteroides succinogenes NCDO 2212
Bacteroides ruminicola subs. brevis NCDO 2212
Bacteroides ruminicola subs. ruminicola NCDO 2202
Butyrivibrio fibrisolvens NCRL
Fusobacterium polysaccharolyticum NCRL

Table 3. Enrichment Medium for Lignocellulose Substrates

Trypticase	5.0 g/L
$(NH_4)_2SO_4$	0.9 g/L
KH_2PO_4	0.9 g/L
NaCl	0.9 g/L
$CaCl_2$	0.02 g/L
$MgCl_2 \cdot 6\ H_2O$	0.02 g/L
$MnCl_2 \cdot 4\ H_2O$	0.01 g/L
$FeSO_4 \cdot 7\ H_2O$	0.1 g/L
Resarzurin	0.33 g/L
Rumen fluid	0.1 cc
Acetic acid	1.7 cc
pH adjusted to 6.5	

The lignocellulosic residues were used as sole carbon source

The alternate process being studied at ICAITI is the conversion of the holocellulose to volatile organic acids using mixed flora. In this case, again the acids accumulate in the medium and are later transformed to ethanol.

Lignocellulose → Organic acids (acetic, lactic)

Organic acids → Ethanol and microbial biomass

The conversion of the lignocellulose to organic acids takes place as a non-sterile anaerobic process. Mixed cultures are used from rumen, marshes, thermophilic fumarola muds, and from the acidogenic portion of a two-phase methane digestor. At the same time, using a synthetic medium (Table 4), and continuous culture techniques, we are screening microorganisms capable of transforming organic acids such as acetic, propionic, and butyric acids to ethanol (4).

Table 4. Synthetic Medium for Conversion of Organic Acids

Yeast extract	15 g/L
Ammonium sulfate	40 g/L
Potassium phosphate, monobasic	1 g/L
Sodium chloride	1 g/L
Calcium chloride	1 g/L
Magnesium sulfate	5 g/L

Organic acids are added to this medium, alone or in combination, at the following levels: acetic acid, 2.8 mL/L; propionic acid, 1 mL/L; and/or butyric acid, 1 mL/L.

CONCLUSION

Research with elective mixed culture without aseptic technique should lead quickly to practical processes. Pure culture studies help to explain the results and can provide a basis for supplementing the systems with special strains should this become desirable.

ACKNOWLEDGEMENT

This study is being carried out on a grant from the Science and Technology Program of US-AID.

REFERENCES

1. Rolz, C., S. de Cabrera and R. Garcia(1979). Ethanol from sugar cane: EX-FERM concept., Biotechnol. Bioeng., 21, 2347-2349

2. Marten, G.C. and R.F. Barnes (1980). Prediction of energy digestibility of forages with *in vitro* rumen fermentation and fungal enzyme systems. In: Standardization of Analytical Methodology for Feeds (Pigden, Balch and Graham, Eds.) IDRC-134e International Development Research Centre, Ottawa, Canada, 61-71

3. Dowman, M.G. and F.C. Collins (1982). The use of enzyme to predict the digestibility of animal feeds. J. Sci. Food Agric., 33, 689.

4. Ohyama, Y. and S. Hara (1975). Growth of yeasts isolated from silages on various media and its relationship to aerobic deterioration of silage. Jap. J. Zootech. Sci., 46 (12), 713-721.

pH INHIBITION OF YEAST ETHANOL FERMENTATION IN CONTINUOUS CULTURE

Robert V. Parsons, Norton G. McDuffie*, George A. Din[1]

Department of Chemical and Petroleum Engineering
and
(1) Department of Biology,
University of Calgary, Calgary, Alberta, Canada T2N 1N4

SUMMARY

In a low dilution rate study, an unexpected pH-related inhibition of yeast fermentation was found. A higher volumetric rate of ethanol production occurred at lower pH values (2.8 to 3.2), suggesting a low optimum pH.

INTRODUCTION

A simple chemostat system was constructed to investigate low dilution rate fermentation behaviour. No direct pH control was employed within the reactor. Instead, similar to the method employed by Jin et al. (1981), the pH of the input medium was set to elevated levels and the system was allowed to come to steady state.

The input pH was approximately 6.6 in initial runs, but was reduced slightly to 6.0 in later runs. This change was intended to induce a slightly lower output pH. No significant changes in output results were anticipated since pH, within a reasonable range, is not usually considered to exert a strong effect on fermentation rates [Jones et al. (1981)].

METHODS AND MATERIALS

Yeast Culture. The yeast strain employed was *Saccharomyces cerevisiae* National Strain Y174, obtained from Alberta Distillers Ltd., Calgary, Alberta.

Chemostat System. A chemostat with a working volume of 210 to 240 mL was fed fresh media using a reciprocating pump at dilution rates ranging from 0.04 to 0.05 h^{-1}. Discharge was removed using a faster peristaltic pump.

Media. The feed media composition was similar to that of Aiba et al. (1968). The composition per liter prior to autoclaving was as follows: Glucose, 45 to 180 g; KH_2PO_4, 6.0 to 7.2 g; $(NH_4)_2SO_4$, 2.0 g; $MgSO_4 \cdot 7H_2O$, 0.4 g Yeast Extract (Difco), 2.0 to 2.4 g.

Analytical Techniques. Reactor output was sampled at 12-hour intervals. Cell concentration was determined by optical density. Glucose concentration was determined using Nelson reagent [Clark (1964)]. Ethanol concentration was determined by gas chromatography (FID) as described in Lie et al. (1970) with acetone as the internal standard. The pH of samples was measured in air using a Radiometer pH meter.

RESULTS AND DISCUSSION

As a result of the slight change in feed pH described in the introduction, pronounced differences were observed in the nature of steady states achieved at similar dilution rates. The earlier runs exhibited lower cell concentrations, lower volumetric rates, and higher specific rates of ethanol formation and substrate consumption than later runs (Table 1). The changes observed in the later runs coincided with a distinct decrease in the output pH. The observed differences thus appeared to be linked to the pH of the system, with an apparent inhibition of metabolism at the higher output pH.

Table 1. Differences in Output Results

Input pH	6.6 Inhibited	6.0 Non-Inhibited
Output pH	5.0 to 5.5	2.8 to 3.2
Cell Concentration (mg/mL)	2.4 to 2.6	5.0 to 8.0
Volumetric Ethanol Production Rate (mg/mL h)	1.0 to 1.5	2.0 to 2.4

The volumetric rates of ethanol formation, ranging overall from 1.0 to 2.4 mg/mL h, were similar to those observed by Cysewski and Wilke (1976) at low dilution rates. However, both the volumetric rates of product formation and substrate consumption were lower at the higher output pH conditions than during the later runs. It was also found that total glucose consumption was adversely affected in the earlier runs (Figure 1), with the fraction of glucose-consumed dropping much more rapidly as the total concentration of the substrate was increased.

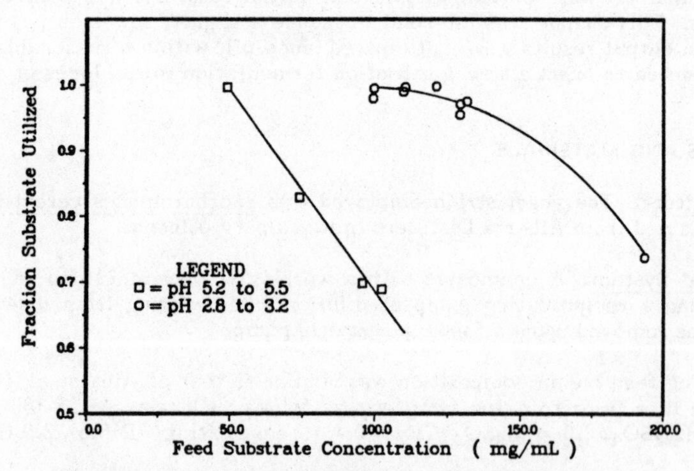

Figure 1. Effect of pH inhibition on the Consumption of Feed Substrate at Steady Conditions

The effect on the value of μ is illustrated in Figure 2, where the following negative exponential ethanol inhibition model [Aiba et al. (1968)] was assumed, using a value of Ks estimated from data to be 0.15 mg/mL:

$$\mu = \mu_0 \, e^{-KiP} \frac{S}{S + Ks}$$

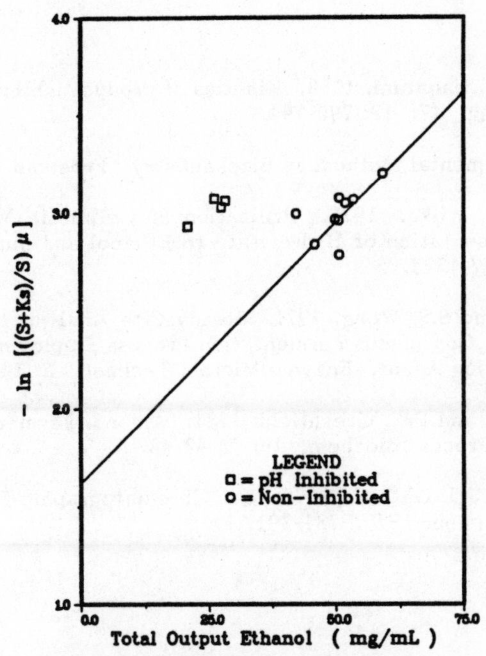

Figure 2. Estimation of Exponential Inhibition Constants – Ks = 0.15 mg/mL

Sufficient points from non-inhibited runs were available to estimate the parameters μ_0 and Ki (Table 2). A model for the system at low pH values (2.8 to 3.2) was derived as follows:

$$\mu = 0.196 \, e^{-0,026P} \frac{S}{S + 0.15}$$

Table 2. Estimates of Exponential Model Parameters from Linear Regression for the Non-Inhibited Case in Figure 2. [ks = 0.15 mg/mL]

Intercept	$-\ln\mu_0$	1.63
Slope	Ki (mg/mL)	0.026

There were not enough data points over a sufficient range of product or substrate concentrations to allow estimation of model parameters for the inhibited (higher pH) case. A definite grouping of the higher-pH data points was, however, apparent in Figure 2, with μ values roughly one-half of those predicted by the non-inhibited model parameters.

No attempt was made to explain the inhibition phenomenon, but the results do have some interesting ramifications. At similar rates of multiplication, the yeast appeared to tolerate higher ethanol concentrations at the lower pH (non-inhibited) condition. Although this behaviour may be unique to the system studied, it does suggest a lower optimum pH for continuous fermentation at low dilution rates, much lower than the pH value of 4.0 generally employed. The results may also apply to cell recycle systems, suggesting that enhanced volumetric ethanol production rates could be achieved using a lower operating pH.

REFERENCES

Aiba, S., M. Shoda and M. Nagatani. 1968. Kinetics of product inhibition in alcohol fermentation. Biotechnol. Bioeng. 47: 12: 790-794.

Clark, J.M. 1964. Experimental Methods in Biochemistry. Freeman, San Francisco. 12-14.

Cysewski, G.R. and C.R. Wilke. 1976. Utilization of Cellulosic Material through Enzymatic Hydrolysis I. Fermentation of Hydrolysate to Ethanol and Single-Cell Protein. Biotechnol. Bioeng. 18: 1297-1313.

Jin, C.K., H.L. Chiang and S.S. Wong. 1971. Steadystate Analysis of the Enhancement in Ethanol Productivity of a Continuous Fermentation Process Employing a Protein-Phospholipid Complex as a Protecting Agent. Enzyme Microb. Technol. 3: 249-257.

Jones, R.P., N. Pamment and P.F. Greenfield. 1981. Alcohol Fermentation by Yeast - The Effect of Environment. Process Biochem. 16: 3: 42-48.

Lie, S., A.D. Haukeli and J.J. Gether. 1970. Gas Chromatographic Determination of Ethanol. Brygmesteren. November 1970. 281-291.

ACID PRODUCTION FROM INSOLUBLE CARBOHYDRATES BY ANAEROBIC DIGESTION

J.M Scharer, V. Devlesaver, P. Girard and M. Moo-Young

Department of Chemical Engineering, University of Waterloo
Waterloo, Ontario Canada N2L 3G1

Organic acids production from cellulose by anaerobic bacterial cultures was investigated. The variables included cellulose concentration (0.5 to 2.3%), carbon to nitrogen ratio (4:1 to 24:1), loading rate, and mixing. Optimum acid production occurred at low carbon to nitrogen ratios. The ammonium ion was important to maintain favourable pH conditions. The loading rate was shown to affect acid distribution. High loading rates (2.3 g/L•day) favoured butyrate, while at low loadings (0.5 g/L•day) acetate and propionate were predominant. Agitation ranged from continuous mixing to daily turnover of the digestion mixtures. The utilization of cellulose was relatively unaffected by the mixing conditions. Organic acid production was not inhibited by ammonium ion concentrations as high as 0.4 M.

INTRODUCTION

Anaerobic digestion of organic matter has become very attractive as a means of stabilizing highly concentrated wastes (industrial and agricultural residues). The primary advantages of the anaerobic process include a higher degree of digestion of lignocellulosic materials at high loading rates (Pfeffer, 1968), reduction of pollution in terms of COD or BOD, production of methane gas and recovery of residues as fertilizer. Currently, anaerobic digestion is viewed as a three-phase process (Taylor, 1982). The first phase involves acidogenic bacteria which hydrolyze and ferment carbohydrates, proteins and lipids to alcohols, volatile fatty acids, H_2 and CO_2. The second phase involves acetogenic bacteria which produce acetate, CO_2 and H_2 from the alcohols and higher fatty acids. Finally, in the third phase, methanogenic bacteria utilize the products of the previous phases, mainly acetate, CO_2 and H_2, to produce CH_4 and CO_2. Because of the different nutritional and environmental requirements of the microbial groups, stage separation seems to be a rational approach in order to optimize the process (Ghosh and Klass, 1978).

Research performed in our laboratory focused on acid production from cellulose. Volatile fatty acid production and acid distribution were assessed as functions of the cellulose loading rate, carbon to nitrogen ratio, agitation conditions and ammonium ion concentration in suspension cultures.

MATERIALS AND METHODS

Organic Acid Production in Suspension Culture

The experimental studies were performed in 500 mL bioreactors and 30 L digester vessels. The 30L apparatus, shown on Figure 1, consisted of an upright, 23 cm diameter, glass cylinder equipped with a conical bottom, recirculation pump, temperature controller, and wet test meter. Cellulose (α-floc) was used as the primary carbonaceous substrate and urea as the nitrogen source. The concentrations of other organic and inorganic constituents were based on levels found in normal swine manure (Lapp, 1977). The inoculum was a mixed culture of bacteria from a fed-batch system adapted to the cellulose substrate. The fermentations were carried out at 39°C. The pH was adjusted to an initial value of 6.0 with HCl. These studies included both batch and fed-batch fermentations with daily addition of fresh medium.

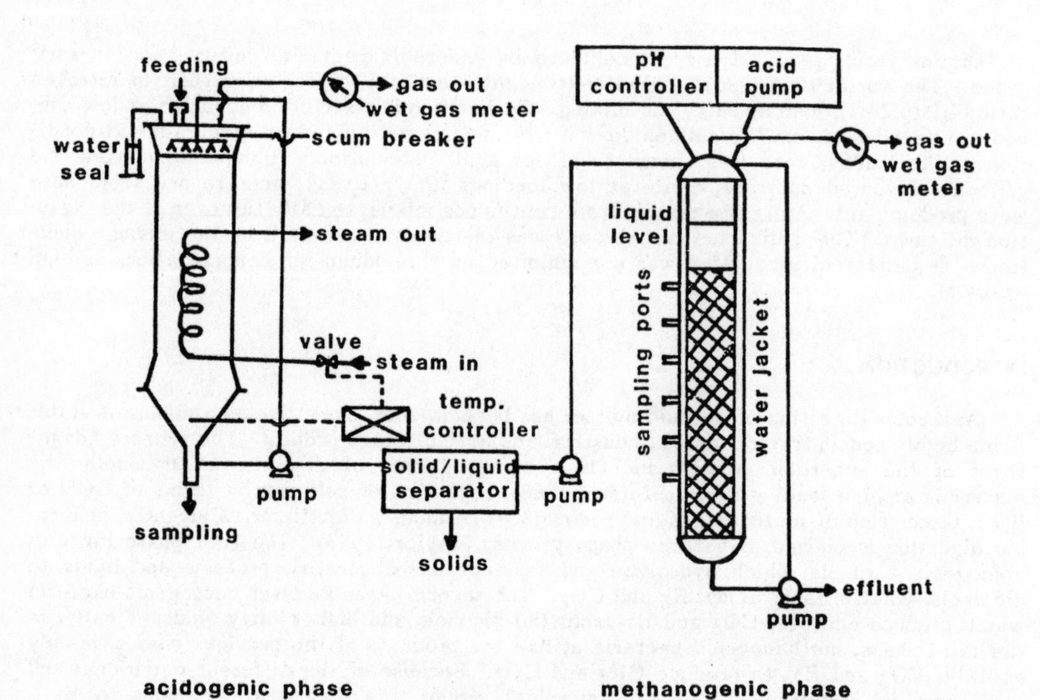

Figure 1: Schematics of the Two-Stage Anaerobic Digestion Apparatus

Organic Acid Utilization and Methane Generation

Analytical techniques included cellulose determinations by a modified Updegraff (1969) method, ammonia measurements by specific ion electrode, carbon dioxide evolution by adsorption in 0.1 N $Ba(OH)_2$ solution (Vogel, 1962), particulate organic nitrogen by the Kjelfoss method, organic acids and biogas analysis by gas chromatography (Scharer et al., 1982).

RESULTS AND DISCUSSION

Organic Acid Production

Fermentations of cellulose to organic acids were performed at carbon to nitrogen (C:N) ratios ranging from 4:1 to 24:1. In Table 1, the product yields, expressed as g acetic acid equivalent/g cellulose added, are compared at low C:N (4:1) and high C:N (24:1) ratios at various cellulose concentrations in the fermentation broth. At low cellulose content in the medium, both the cellulose utilization efficiency and the product yield were essentially independent of the C:N ratio. At high cellulose concentration, however, both the cellulose utilization effiecencey and the product yield declined significantly as the C:N ratio increased. This decline is believed to result from the depressed pH conditions experienced with higher C:N ratios. During the fermentation, the primary nitrogen source, urea, was readily hydrolyzed to ammonia and carbon dioxide. At high C:N ratios coupled with high cellulose levels, the ammonium ion product was insufficient to poise the pH (i.e., neutralize acids) in the optimum range for cellulose hydrolysis and acid generation.

Table 1. Effect of Carbon to Nitrogen Ratio on Volatile Acid Yield

Initial cellulose content (%)	C:N ratio	Fermentation time (days)	Cellulose utilization	Final pH	Yield (g acetic acid equiv./ g cellulose
0.5	4:1	5	82.1	6.1	0.57
0.55	24:1	5	87.3	5.8	0.47
1.0	4:1	7	93.5	6.1	0.75
1.0	24:1	7	79.8	5.5	0.32
2.3	4:1	11	92.8	6.2	0.87
2.3	24:1	11	68.4	5.0	0.45

The effect of cellulose loading rate on acid production and acid distribution at a C:N ratio of 4:1 is shown in Table 2. At a constant hydraulic retention time (5 days), the cellulose utilization efficiency declined with increasing loading rate. The increase of the retention time from 5 days to 11 days, however, was beneficial with regard to the cellulose utilization efficiency. As the loading rate increased, butyric and higher carbon fatty acids (valeric, caproic acids) became progressively more prominent in the fermentor effluent.

Table 2. Effect of Cellulose Loading Rate on Cellulose Utilization and Acid Distribution

Loading rate (g/L•day)	Retention time (days)	Acid Distribution			Cellulose Utilization (%)
		acetate (M)	propionate (M)	butyrate (M)	
0.5	5	0.0250	0.0026	0.0014	82.1
1.0	5	0.0410	0.0183	0.0028	72.6
2.3	5	0.0480	0.0196	0.0026	37.4
2.3	11	0.1030	0.0330	0.0267	92.2

Effect of Agitation

The effect of mixing on acid production was investigated under 3 conditions: continuous agitation at 150 rpm, intermittent mixing at 5 minutes every 2 hours and daily turnover. The fermentations were performed with 1% cellulose at C:N ratio of 4:1. The percentage of cellulose utilization was measured daily and plotted as a function of time. Figure 2 shows that the degree of agitation used in this study has no significant effect on cellulose utilization. A daily mixing of the fermentor contents appears to be adequate.

Ammonium Ion Concentration

In Table 3, acid production from cellulose is shown at high C:N (carbon to nitrogen) and at low C:N ratios. The ammonium ion resulted from the hydrolysis of the urea feed. Since the loading rate of cellulose was held constant, high C:N ratios gave lower concentrations of ammonia in solution. Because of acid production at low ammonia concentrations (0.012 M - 0.025 M NH_4^+), pH was depressed. Low pH conditions, in turn, adversely affected the rate of acid production. Low C:N ratios resulted in higher ammonium concentrations (0.22 M - 0.4 M) hence, higher pH values. The beneficial effect of near neutral pH values on acid production is evident. It is particularly interesting that acetic acid production is improved at the higher pH values. More than four-fold increase in acetic acid concentrations were observed at lower C:N ratios (higher pH values). Increases in propionic and butyric acid content were usually less dramatic.

Figure 2: Cellulose Utilization vs Time at 1% (w/v basis) C/N = 4/1

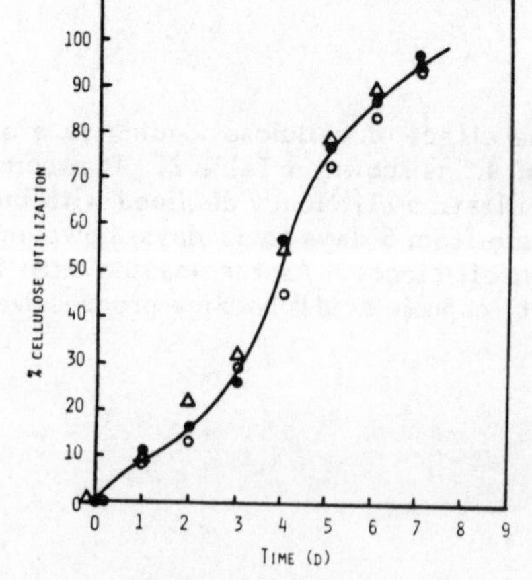

Table 3. Effect of Ammonium Ion on Acid Production from Cellulose

pH	C:N ratio	Acetic (g/L)	Propionic (g/L)	Butyric (g/L)	NH_4^+ (g/L)
5.64	24:1	0.73	2.05	0.26	0.20
6.02	24:1	1.14	1.79	0.14	0.40
5.65	24:1	1.21	1.80	0.11	0.43
7.30	4:1	2.97	6.17	0.26	3.79
7.30	4:1	5.12	3.55	0.48	4.96
7.44	4:1	5.20	1.61	0.21	6.67

Loading rate = 4 g cellulose/L•day

REFERENCES

Ghosh, S. and D.L. Klass (1978). Two phase anaerobic digestion, Process Biochem., 13, 15-24.

Lapp, H.M. (1977). A study of the feasibility of using methane gas produced from animal waste for energy purposes, Agriculture Canada.

Pfeffer, J.T. (1968). Increased bondings on digester with recycle of digested solids, Journal WPCF, 40 (11), 1920-1933.

Scharer, J.M., M. Fujita and M. Moo-Young (1982). Methane generation of agricultural wastes, Waste Treatment and Utilization, Theory and Practice of Waste Management, Vol. 2 by M. Moo-Young et al., Pergamon Press.

Taylor, G.T. (1982). The methanogenic bacteria, Progress in Industrial Microbiology, Vol. 16, 231-329.

Updegraff, D.M. (1969). Semimicro determination of cellulose in biological materials, Anal. Biochem., 32, 420-424.

Vogel, A.I. (1962). A Textbook of Quantitative Inorganic Analyses, 3rd Edition, Longman.

DEVELOPMENTS IN METHANOGENIC REACTOR DESIGN

L. van den Berg

Division of Biological Sciences
National Research Council of Canada
Ottawa, Ontario CANADA K1A 0R6

INTRODUCTION

Industrialization and urbanization, with the concomitant serious waste disposal problems, resulted in widespread practise of methane recovery from primary and activated sludge (van Brakel, 1980). Process improvements in anaerobic digestion, however, were few and far between from the turn of the century to about 1950. It was recognized that temperature affected the process drastically and that mixing was beneficial. Temperature control (using the methane produced as a source of energy) and mixing were therefore introduced over the years. Process control improved with the observation that organic (volatile) acids accumulated under sub-optimum conditions and that the amount and the composition of the gas produced were important indicators of the health of reactors. Major improvements, however, depended on a more thorough understanding of the process. Unfortunately, the extreme anaerobic requirements of the bacteria involved proved a stumbling block until better methods for studying anaerobes became available after 1950 (Hungate, 1950).

Of major significance in the development of better anaerobic methanogenic reactors was the realization that much of the performance of anaerobic digestors depended on the conversion of acetic acid to methane and that the bacteria involved grew very slowly (mass doubling times of 8-80 days under practical conditions). This agreed with early empirical observations that hydraulic residence times in excess of 15 days were required for satisfactory operation. The newer types of reactors succeeded in extending the retention time of the bacteria in the reactor (to usually well over 15 days) while at the same time reducing hydraulic residence times (reactor volume divided by daily volumetric flow) to a few hours or a few days. The concentration of bacteria in the reactor was increased at the same time, leading to markedly increased rates of conversion of organic material to methane.

METHANE PRODUCTION

The rate of methane production in a reactor is proportional to the concentration of organic material in the substrate solution, and the fraction of this material that is converted to methane. This rate is also inversely proportional to the hydraulic residence time and the oxygen content of the organic material converted to methane. For example, carbohydrates yield less methane than fats (0.3 and up to 1 m^3/kg, respectively).

Many stirred tank reactors (e.g., municipal digestors) have volumetric methane production rates of only about 0.5 m^3/m^3/day with hydraulic residence times of over 20 days. The stoichiometry of the conversion reactions indicate that these rates of methane production are only possible with substrate concentrations of at least 35 kg/m^3. To treat more dilute wastes, even at these low methane production rates, it is necessary to reduce hydraulic residence times. As mentioned, this is possible in advanced type reactors in which active biomass is retained in high concentrations, in spite of short hydraulic residence times. In addition, these advanced reactor types can have methane production rates of over 5 m^3/m^3/day at hydraulic residence times of less than half a day.

CHARACTERISTICS OF ADVANCED REACTORS

This paper is limited to a discussion of the five most widely studied and used methanogenic, retained-biomass reactors. Retained-biomass reactors for anaerobic waste treatment were first systematically studied with the development of the anaerobic contact process in the fifties. The other process was followed by the development of the anaerobic filter. The other processes (upflow anaerobic sludge bed reactors, fluidized or expanded bed reactors and downflow stationary fixed film reactors) were developed in the seventies. Recently, two-phase reactors have been developed based on recent insights into the underlying microbiological and biochemical processes involved in methanogenesis, but these are not discussed here. More details are available in other recent reviews (Kirsop, 1984; van den Berg, 1983).

Anaerobic Contact Reactor

This was the first retained-biomass reactor to be studied and developed systematically (Schroepfer et al., 1955). The principle involved is the same as in the activated sludge process -settling of microbiological floc and other suspended solids and contacting the raw waste with the anaerobic sludge (Fig. 1a). The reactor's performance depends markedly on the efficiency with which the microorganisms and suspended solids settle. An additional factor is the degree to which sludge and raw waste are mixed in the reactor. Good mixing is required, yet it is essential that the settling characteristics are not adversely affected. The process is especially suited for wastes with a certain amount of hard-to-digest solids that settle readily or attach themselves readily to solids which settle readily.

Major limitations of the process are caused by the difficulties of obtaining good settling (Steffen, 1961; van den Berg and Lentz, 1978) and, in large reactors, of providing adequate mixing. Generally, up to 80% of the microorganisms may settle, indicating that the hydraulic retention time cannot be less than one fifth of the minimum mass doubling time. The latter usually is 10 days or more. For many wastes, the settling efficiency is less than 80% and the minimum hydraulic residence times are correspondingly longer.

The anaerobic contact process has been used commercially in several countries including the U.S.A., Sweden, France and Canada. It is especially useful for wastes containing finely dispersed organic matter, such as starch.

Anaerobic Filter

This reactor, which contains a solid support or packing material, was developed by Young and McCarty (1967). Waste is added at the bottom and flows upward (Fig. 1b). The packing itself serves to separate the gas and to provide quiescent areas for settling of suspended growth. Bacteria are retained mostly in suspended form with a relatively small portion attached to the surface (Dahab and Young, 1982; van den Berg and Lentz, 1979). Most of the bioconversion occurs near the bottom of the reactor since the suspended biomass tends to collect near the bottom. The growth on the surfaces of the packing provides a polishing action (Dahab and Young, 1982; Young and Dahab, 1982). The process is particularly suitable for dilute soluble wastes, soluble wastes which can be made dilute by recirculating effluent or wastes with easily degradable suspended material.

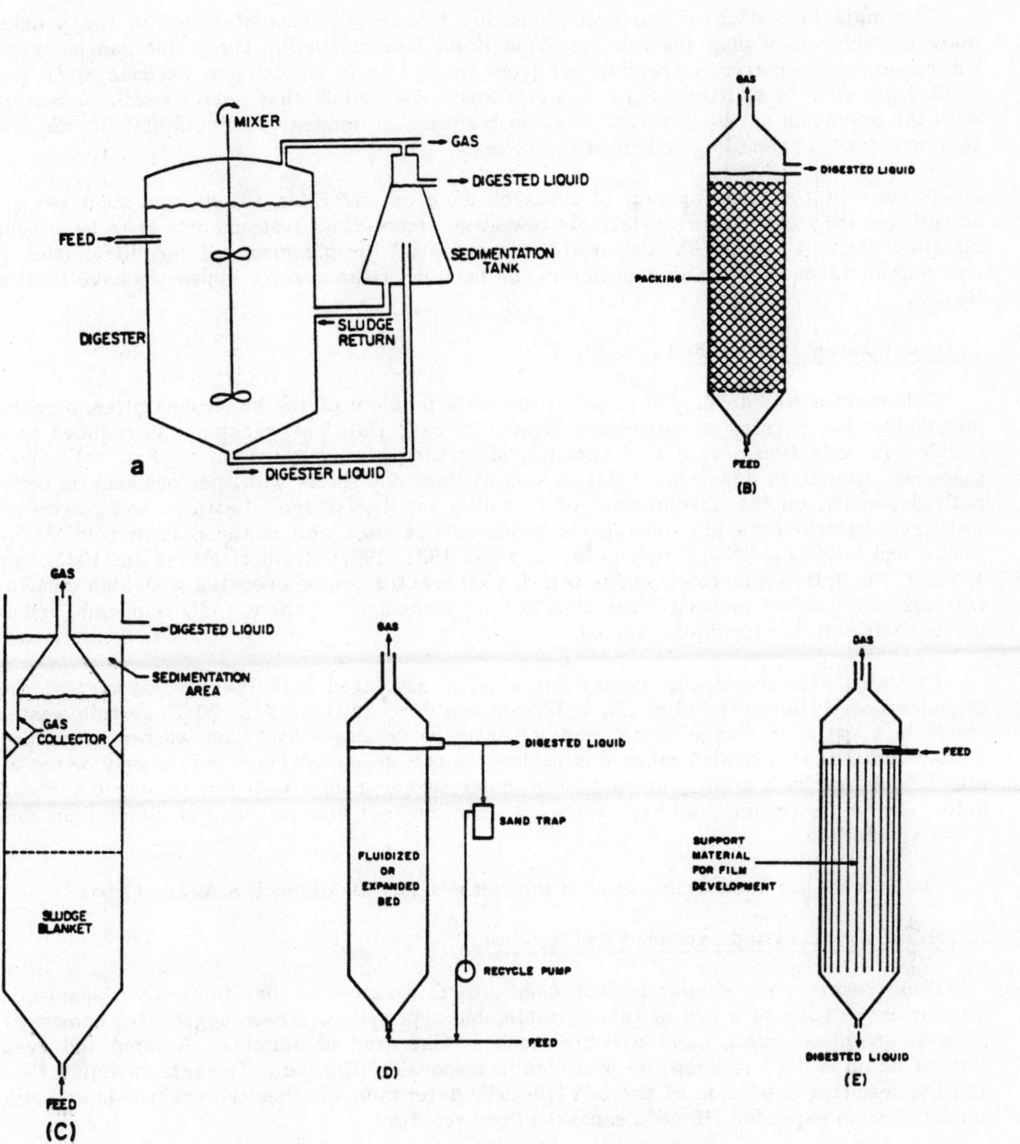

Fig. 1. Sketches of the anaerobic contact (a), anaerobic filter (b), upflow anaerobic sludge bed (c), anaerobic fluidized or expanded bed (d), and downflow stationary fixed film (e) reactors.

The main limitation of the process is due to accumulation of solids in the packing material which may plug the reactor (Young and Dahab, 1982). The solids can be waste suspended solids, material precipitated from the waste (e.g., calcium carbonate) or suspended growth. In addition, hard-to-digest suspended solids that settle readily interfere with the operation of the reactor. In large reactors, an inadequate liquid distribution system may cause channelling and short circuiting.

In spite of the large amount of research done on anaerobic filters, and their several advantages for many wastes, relatively few large commercial systems appear to have been installed (Witt et al., 1979). Generally, the potential for plugging and the difficulties of ensuring an adequate flow distribution in the bottom of the reactor appear to have limited its use.

Upflow Anaerobic Sludge Bed Reactor

This reactor was developed to avoid the main problem of the anaerobic filter, namely, plugging of the packing by suspended growth of bacteria. The packing was reduced to a simple gas collection device that encouraged settling of suspended solids (Fig. 1c). Since suspended growth in anaerobic reactors usually does not settle well, performance is critically dependent on the development of a readily settling sludge. Lettinga and coworkers were able to develop a granular sludge which settles very well in the reactor (Fig. 2) (de Zeeuw and Lettinga, 1980; Lettinga et al., 1980, 1981, 1982; Hulshoff Pol et al., 1982). As a result, the upflow anaerobic sludge bed (UASB) reactor can be operated with high concentrations of microbial biomass, very high loading rates and excellent COD removals, and is particularly suitable for dilute wastes.

Problems with the UASB reactor are usually associated with the development of the granular sludge (Hulshoff Pol et al., 1982; van den Berg et al., 1981). While certain wastes result in a granular sludge quite readily (sugar processing waste and wastes containing mainly volatile fatty acids), other wastes develop this granular sludge very slowly or not at all. Inoculation with a large amount of granular sludge from a well functioning UASB can help. The sludge retains its characteristics often, but not always, when changing from one waste to another.

The process has been widely used in the Netherlands, Belgium, U.S.A. and Cuba.

Anaerobic Fluidized and Expanded Bed Reactors

These reactors are similar to suspended growth reactors in that the active biomass is present in the form of a bed of readily settleable aggregates. These aggregates consist of biomass grown on small, inert particles such as fine sand or alumina. A rapid and even flow of liquid is used to keep the particles in suspension (Fig. 1d). The rate of liquid flow and the resulting expansion of the bed (10-25%) determine whether the reactor is called a fluidized or an expanded (10-15% expansion) bed reactor.

The preferred waste substrate for these reactors is soluble or at least the suspended material should be easily degradable (Switzenbaum, 1982). Reactor performance depends very much on the evenness of the flow of the liquid and as a result, the system of liquid distribution is very critical (Jewell, 1982). The capital cost of the flow distribution system and the pumps is high, while the net energy yield is lower than for other reactors. The first large scale anaerobic fluidized and expanded bed reactors are under construction in the U.S.A.

Downstream Stationary Fixed Film Reactor

This reactor was also developed to avoid the plugging problems of the anaerobic filter. The reactor contains packing and is operated in the downflow mode (Fig. 1e). Suspended growth and indigestible, suspended waste solids are removed with the effluent. The need for an elaborate distribution system is also eliminated because waste entering at the top of the reactor is readily dispersed by the gas escaping from the packing (Duff and Kennedy, 1983a). The important factor in this reactor is the formation and stability of an active biomass film on the surfaces provided (Murray and van den Berg, 1981; van den Berg and Kennedy, 1980). To avoid accumulation of non-active suspended material in the packing, the architecture of the packing is important.

The downflow stationary fixed film (DSFF) reactor is capable of handling a wide variety of wastes, from reasonably dilute to very concentrated ones (Kennedy and van den Berg, 1982a, 1982b; van den Berg and Kennedy , 1982). Suspended solids, such as those present in some food processing wastes and in manures, are readily accommodated, although their degradation depends on the time they spend in contact with active biomass. The reactor can operate over a wide range of temperatures (Duff and Kennedy, 1983b; Kennedy and van den Berg, 1982c, d).

Loading rates are limited by the amount of active biomass that can be retained in the reactor (Kennedy and van den Berg, 1982b). Effective film thickness is limited by diffusion; hence, the amount of biomass is a function of the surface area available for film formation. This area is limited to less than 100 m^2/m^3 because the channels in the packing have to be large enough to prevent filling up with film. Another limitation is on the use of very dilute waste. To obtain reasonable high loading rates with dilute wastes, the hydraulic retention time has to be short. Since the channels have a minimum dimension, the probability of contact between waste organics and film decreases with decreasing hydraulic retention time.

The process has been applied commercially in Puerto Rico and in Canada.

FUTURE DEVELOPMENTS

The reasons for the use of anaerobic treatment with methane recovery rather than aerobic treatment for waste waters are compelling: net energy production instead of energy consumption, much higher loading rates, simpler operation and reduced capital costs. Since anaerobic methane production is not a complete waste treatment system, and its effluent needs to be treated further, it is necessary to reverse present treatment philosophy: where at all possible, anaerobic treatment should precede, rather than follow, aerobic treatment.

Further improvements in efficiency and rates of production, and reduction in costs of operation are possible in the near future. Basic microbiological research in Canada and elsewhere will undoubtedly lead to better control systems, faster rates of reactions and higher conversion efficiencies (e.g., by faster hydrolysis of cellulose, proteins and fats or by choosing optimum conditions for what are now largely unknown microoganisms). Work with cocultures (Laube and Martin, 1983) is starting to show promising relationships between different types of bacteria in anaerobic reactors.

It is possible to improve the performance of the advanced reactors further, or modify them to become cheaper to install and simpler to operate. The rapidly developing understanding of the microbial and biochemical processes involved, especially those involving ecological relationships, will be of great value in obtaining further improvements. The full potential of the sludge bed and expanded bed reactors for the treatment of raw sewage, for example, has not been reached as yet. Combinations of the reactor types discussed here are being researched aggressively. In addition, it can be expected that other methods of cell retention will undoubtedly be invented and developed (Bachmann et al., 1982; Binot et al., 1982; Guiot and van den Berg, 1984; Martensson and Frostell, 1982; Oleszkiewicz and Olthoff, 1982).

Applied research in Canada is aimed at more efficient and cheaper reactor designs, and automatic and less expensive reactor operation. Thus a side-by-side comparison of several advanced reactors is important. This is done by Environment Canada in cooperation with the National Research Council, in a pilot plant facility (Hall, 1982; Hall et al., 1982). On the other end of the scale (slow rates of methane production), a Canadian firm has had commercial success in upgrading an old anaerobic lagoon system by installing methane recovery systems for their "bulk volume reactor" (Landine et al., 1981).

The practical developments so far have provided waste treatment engineers with effective and inexpensive methods of waste treatment which undoubtedly will find increasing use in practise. These developments have also opened up a fascinating field for microbiological studies.

REFERENCES

Bachmann, A., V.L. Beard, and P.L. McCarty. 1982. Comparison of fixed film reactors with a modified sludge blanket reactor. Proc. 1st Int. Conf. Fixed Film Biol. Processes, 1192-1211.

Binot, R.A., T. Bol, H.T. Naveau and E.J. Nyns. 1983. Biomethanation by immobilized fluidized cells. Water Sci. Technol., 15, 103-116.

Dahab, M.F., and J.C. Young. 1982. Retention and distribution of biological solids in fixed-bed anaerobic filters. Proc. 1st Int. Conf. Fixed Film Biol. Processes, 1337-1351.

de Zeeuw, W., and G. Lettinga. 1980. Acclimation of digested sewage sludge during start-up of an upflow anaerobic sludge blanket (UASB) reactor. Proc. 35th Purdue Indust. Waste Conf., 39-47.

Duff, S.J.B., and K.J. Kennedy. 1983a. Effect of effluent recirculation on start-up and steady-state operation of the downflow stationary fixed film (DSFF) reactor. Biotechnol. Letts., 5(5), 317-320.

Duff, S.J.B., and K.J. Kennedy. 1983b. Effect of hydraulic and organic overloading on thermophilic downflow stationary fixed film (DSFF) reactor. Biotechnol. Letts., 4(12), 815-820

Guiot, S.R., and L. van den Berg. 1984. Performance and biomass retention of an upflow anaerobic reactor combining a sludge blanket and a filter. Biotechnol. Letts., 6(3), 161-164.

Hall, E.R. 1982. Biomass retention and mixing characteristics in fixed film and suspended growth anaerobic reactors. Proc. IAWPR Seminar Anaerobic Treatment of Waste Water in Fixed Film Reactors. Copenhagen.

Hall, E.R., M. Jovanovic and M. Pejic. 1982. Pilot studies of methane production in fixed film and sludge blanket anaerobic reactors. Proc. 4th Bioenergy R&D Seminar, Winnipeg, 475-479.

Hulshoff Pol, L.W., W.J. de Zeeuw, C.T.M. Velzeboer and G. Lettinga. 1982. Granulation in UASB reactors. Water Sci. Technol., 15, 291-304.

Hungate, R.E. 1950. The anaerobic mesophilic cellulolytic bacteria. Bacterial. Reviews, 14, 1-49.

Jewell, W.J. 1982. Anaerobic attached film expanded bed fundamentals. Proc. 1st Int. Conf. Fixed Film Biol. Processes, 17-42.

Kennedy, K.J., and L. van den Berg. 1982a. Anaerobic digestion of piggery-waste using a stationary fixed film reactor. Agr. Wastes, 4, 151-158.

Kennedy, K.J., and L. van den Berg. 1982b. Effect of height on the performance of anaerobic downflow stationary fixed film (DSFF) reactors treating bean blanching waste. Proc. 37th Purdue Indust. Waste Conf., 71-76.

Kennedy, K.J., and L. van den Berg. 1982c. Stability and performance of anaerobic fixed film reactors during hydraulic overloading at 10-35 °C. Water Res., 16, 1391-1398.

Kennedy, K.J., and L. van den Berg. 1982d. Thermophilic downflow stationary fixed film reactors for methane production from bean blanching waste. Biotechnol. Letts., 4(3), 171-176.

Kirsop, B.M. 1984. Methanogenesis. Crit. Rev. Biotechnol., 1, 109-159.

Landine, R.C., A.A. Cocci, T. Viraraghavan, and G.J. Brown. 1981. Anaerobic pretreatment of potato processing waste water -a case history. Proc. 36th Purdue Indust. Waste Conf., 233-240.

Laube, V.M., and S.M Martin. 1983. Effect of some physical and chemical parameters on the fermentation of cellulose to methane by a coculture system. Can. J. Microbiol., 29(11), 1475-1480.

Lettinga, G., W. de Zeeuw, and E. Ouborg. 1981. Anaerobic treatment of wastes containing methanol and higher alcohols. Water Res., 15, 171-182.

Lettinga, G., S.W. van Velson, W. Hobma, W. de Zeeuw and A. Klapwyk. 1980. Use of the upflow sludge blanket (USB) reactor concept for biological waste water treatment especially for anaerobic treatment. Biotechnol. Bioeng., 22, 699-734.

Martensson, L., and B. Frostell. 1982. Anaerobic waste water treatment in a carrier assisted sludge bed reactor. Water Sci. Technol., 15, 233-246.

Murray, W.D., and L. van den Berg. 1981. Effect of support material on the development of microbial fixed films convertin acetic acid to methane. J. of Appl. Bact., 51, 257-265.

Oleszkiewicz, J.A., and M. Olthoff. 1982. Anaerobic treatment of food industry waste waters. Food Tech., 78-82.

Schroepfer, G.J., W.J. Fuller, A.S. Johnson, N.R. Ziemke, and J.J. Anderson. 1955. The anaerobic contact process as applied to packinghouse wastes. Sewage Ind. Wastes, 27, 460-486.

Steffen, A.J. 1961. Operating experiences in anaerobic treatment of packinghouse waste. Proc. 3rd Research Conf., Am. Meat Indust. Foundation, 81-89.

Switzenbaum, M.S. 1982. A comparison of the anaerobic filter and the anaerobic expanded/fluidized bed processes. Water Sci. Technol., 15, 345-358.

van Brakel, J. 1980. The Ignis Fatuus of biogas. Delft University Press.

van den Berg, L., and K.J. Kennedy. 1980. Support materials for stationary fixed film reactors for high-rate methanogenic fermentations. Biotechnol. Letts., 3(4), 165-170.

van den Berg, L, and K.J. Kennedy. 1982. Effect of substrate composition on methane production rates of downflow stationary fixed film reactors. Proc. IGT Symp. Energy from Biomass and Wastes, 401-424.

van den Berg, L., and K.J. Kennedy. 1983. Comparison of advanced reactors. Proc. 3rd Intern. Symp. Anaerobic Digestion, Boston, Mass., 71-89.

van den Berg, L., K.J. Kennedy, and M.F. Hamoda. 1981. Effect of type of waste on performance of anaerobic fixed film and upflow sludge bed reactors. Proc. of 36th Purdue Indust. Waste Conf., 686-692.

van den Berg, L., and C.P. Lentz. 1978. Factors affecting sedimentation in the anaerobic contact fermentation using food processing wastes. Proc. 33rd Purdue Indust. Waste Conf., 185-193.

van den Berg, L., and C.P. Lentz. 1979. Comparison between up- and downflow anaerobic fixed film reactors of varying surface-to-volume ratios for the treatment of bean blanching waste. Proc. 34th Purdue Indust. Waste Conf., 319-325.

Witt, E.R., W.J. Humphrey, and T.E. Roberts. 1979. Full-scale anaerobic filter treats high strength wastes. Proc. 34th Purdue Indust. Waste Conf., 229-234.

Young, J.C., and M.F. Dahab. 1982. Effect of media design on the performance of fixed bed anaerobic reactors. Water Sci. Technol., 15, 369-384.

Young, J.C., and P.L. McCarty. 1967. The anaerobic filter for waste treatment. Proc. 22nd Purdue Indust. Waste Conf., 559-574.

A LARGE-SCALE BIOLOGICALLY DERIVED METHANE PROCESS

Donald L. Wise, Alfred P. Leunschner, Mostafa A. Sharaf

Dynatech R/D Company
99 Erie Street
Cambridge, Mass.
U.S.A. 02139

SUMMARY

A low-capital intensive system for the production of methane fuel gas from all available organic residues is described. Application to lesser developed countries (LDCs) is presented in some detail. The low capital cost system proposed is based on recent experience with methane fuel gas recovery from municipal solid waste (MSW) landfill sites via anaerobic digestion and from experiments on batch anaerobic digestion of mixed agriculture residues. Thus, the "process" is simple batch anaerobic digestion; the combined organic residues are intended to be all the available organic residues from a populatous area, i.e., all residues in the local wasteshed.

The advantages of the proposed system to an LDC include:

a) energy recovery from various residues in the form of a useful and acceptable "biogas" fuel not otherwise recoverable;
b) a simple process -- as simple as landfilling of MSW;
c) slow capital cost -- costs associated with the labour of digging a large pit;
d) utilization of all organic residues in a populous area -- this includes
   - all municipal wastes (solid wastes, garbage, sewage sludge)
   - industrial residues such as those from food processing
   - agricultural residues such as wheat, rice, and barley straws as well as animal manures;
e) waste "stabilization", i.e. waste treatment, especially valuable for sewage sludge.

INTRODUCTION

A three phase program, now in progress, is described. First, work has begun in the operation of anaerobic digestors in the laboratories of ten selected LDC institutions around the world. Earlier related experiments at Dynatech have been successful using 55 gallon barrels as batch anaerobic digestion reactors. These simple digesters are useful to contain the residues, and may incorporate simple devices to monitor methane fuel gas production. These simple digesters will also be used with all the international participants in this program. Straight-forward experiments are being initiated at the ten sites around the world. Results will be used to make useful comparisons between methane fuel gas production from combined residues in selected LDCs. The second phase of the proposed program, also now being initiated, will be the preliminary design and cost estimation, on a uniform basis using a computer model for implementing this low capital cost fuel gas production system in selected LDCs. The culmination of the proposed program, phase three, will be a workshop at Dynatech including all participants from around the world. The results of this workshop and recommendations will be presented to the sponsor, the U.S. Agency for International Development (A.I.D.). Of special interest will be plans for practical implementation in LDCs.

WORLD-WIDE PARTICIPANTS

Integral to this program is the current establishment of "linkages" between key workers around the world, who have either direct background in anaerobic digestion or sufficient technical capabilities and interest in this bioenergy conversion project to make a significant contribution. These "linkages" will be useful for acceptance of the results of this joint work and early implementation in LDCs. Thus, by conducting rather simple experiments with workers around the world using local combined residues, it will be possible to learn of the broader world-wide potential for this low capital energy recovery system. In this way, full scale implementation should be most acceptable.

The following is a summary, in alphabetical order, of all international participants.

Country	Principal Investigator(s)
Egypt	Dr. M. Nabil Alaa El-Din National Director of Biogas Project Institute of Soils and Water Research Dept. of Agricultural Microbiology Giza
Guatemala	Dr. Carlos Rolz Head, Applied Research Division Instituto Centroamericano de Investigacio y Tecnologia Industrial Central American Research Institute for Industry Avenida La Reforma 4-47 Zona 10
India	Dr. V.V. Modi Professor and Head Dept. of Microbiology The Maharaja Sayajirao University of Baroda Baroda - 390 002 India
Indonesia	Dr. Didin S. Sastrapradja Deputy Chairman for Natural Sciences Kantor/Office Lembaga Ilmu Pengetahuan Indonesia Jln, Teuku Cik Ditiro 43 Jakarta
Israel	Dr. Uri Marchaim Kibbutz Industries R&D Institute Kibbutz Kfar-Giladi Israel 12-210

Macau/Hong Kong

Dr. Chen Ru-Chen
Manager, Technical Department
Patex Import-Export
13 K Rua De Manuel De Arriaga
Macau via Hong Kong

and

Dr. M.H. Wong
Senior Lecturer
Department of Biology
University Science Centre
The Chinese University of Hong Kong
Shatin. NT
Hong Kong

Malasia

Dr. Kee Kean Chin
National University of Singapore
Faculty of Engineering
12 B Patterson Hill, Singapore 9

Peru

Dr. Javier Verastegui
Jr. Morelli 2da. cda
esquina av. de las Artes (alt. cda.
21 av. Javier Prado-este)
San Forja - Surquillo Lima, 34
apartado 145 Lima

Philippines

Dr. William G. Padolina
Department of Chemistry
University of the Philippines
at Los Banos
College, Laguna 3720

Spain

Dr. Joan Mata
Universidad de Barcelona
Facultad de Quimica
Diagonal 647
Barcelona 28

DYNATECH'S BACKGROUND

Engineers at Dynatech have pioneered the development of practical low cost batch anaerobic digestion systems. Research activities at Dynatech have been directed at the recovery of methane fuel gas from a) landfills of municipal solid waste (including sewage sludge), and b) combined agricultural residues (batch anaerobic digestion of rice, wheat, and barley straws have been evaluated along with dairy and beef cattle manure, as well as food processing wastes such as tomato and onion/garlic wastes). This work is briefly summarized below.

With respect to municipal solid waste and sewage sludge, attention has been given at Dynatech to the concept of a "controlled" landfill in which nutrients, buffer, etc., are added to the combined solid waste sewage prior to landfilling in order to enhance fuel gas production. Attention has also been directed to adding these ingredients to existing landfills.

The work with agricultural residues also included adding to the straws, food processing wastes, such as tomato wastes and animal manures. Thus, a unique feature of the Dynatech batch digestion work has been the emphasis on the use of combined agricultural residues. This use of combined agricultural residues recognizes the practical use of all wastes in an agricultural wasteshed.

It is to be noted that several well respected papers have been published on this work at Dynatech. The first paper describes the "controlled" landfilling of municipal solid wastes (Augenstein et al., 1976). A more recently published paper describes continued work at Dynatech with MSW (Buivid et al., 1980). Another paper describes the *in situ* methane fermentation of combined agricultural residues (Wise et al., 1981). New work, as yet incomplete, includes extension of this original work to three different municipal landfills.

DISCUSSION ON ANAEROBIC DIGESTION

Historically, anaerobic digestion has been used for the treatment of various relatively dilute liquid wastes or suspensions, such as waste from sewage, packing houses, canneries, and sulfite liquors. Its function has been primarily to change the character and reduce the amount of wastes to render them largely inoffensive. Although fuel gas generation has traditionally been of secondary importance, methane fermentation is now viewed with much more interest as a method of producing fuel. Standard digester designs and operating procedures are generally considered optimal for treating fluid wastes. However, when this technology is applied to solid digestible organic substrates, considerable expense and energy consumption accrue from the necessity to grind and slurry the substrate, mix reactor contents continuously, separate solid and liquid components at the end of the digestion process, and dispose of process effluents, particularly the liquids. In recent years, singificant advances have been made in the digestion of solid substrates under minimal moisture conditions. The principles of "dry" anaerobic fermentations have been successfully applied to agricultural residues (Jewell et al., 1981) and municipal solid wastes. This has eliminated the need for slurrying solid substrates, using costly materials of construction (such as steel and concrete), and post digestion materials handling problems.

PRINCIPLES

Methane gas is produced by the destruction of organic material present in solid residue. The term "anaerobic digestion" refers to the ability of certain classes of microorganisms to grow on a number of different organic compounds in the absence of oxygen (hence the term anaerobic), converting them ultimately to the gases carbon dioxide and methane. This process is referred to traditionally as "anaerobic digestion." The technology of anaerobic digestion has long been utilized to dispose of waste materials, especially sewage sludge solids. Methane is generated in anaerobic digestion, and the usefulness of the gas was recognized as the sludge disposal process was developed in the nineteenth century. Since that time, anaerobic digestion has been utilized throughout the world to stabilize sewage solids, and often, the gas produced has been used to provide heat and power for the treatment plant. During the twenty years after World War II, attention to the development of anaerobic digestion lagged as aerobic treatment and tertiary sewage treatment processes were developed, due mainly to their ability to treat sewage more completely, and because of their relatively low costs. As energy costs have escalated rapidly during recent years, however, attention has been focused again on anaerobic digestion as a waste treatment process.

Further expansion of the technology occurred during the late 1960's as anaerobic digestion was viewed with renewed interest as a potential source of energy and as a waste disposal process for various types of organic wastes, including municipal solid waste and agricultural residues. Dr. Clarence G. Golueke, Research Biologist at the University of California (Richmond Field Station), was the first of the more recent workers to study anaerobic digestion of solid waste. A five-year program under his direction examined the technical feasibility of digesting municipal wastes admixed with large quantities of animal wastes (Golueke and McCaunen, 1970; Golueke, 1971). Further work at the University of Illinois, under the direction of Dr. John T. Pfeffer, demonstrated the feasibility of methane production from municipal solid waste, with limited additions of sewage sludge (Pfeffer, 1974; Pfeffer and Liebman, 1974).

In 1969, an intensive laboratory and engineering evaluation of methane fuel gas production from municipal solid waste was initiated at Dynatech R/D Company by Consolidated Natural Gas Service Company, Inc., Cleveland, Ohio. Laboratory experiments conducted at Dynatech reaffirmed the technical feasibility of utilizing the basic anaerobic digestion process to convert organic wastes to pipeline-quality fuel gas. Experimental results demonstrated conclusively that scale-up of this fuel gas production process was merited. Engineers at Dynatech then carried out two programs simultaneously: an experimental development of new concepts for economic production of fuel gas from solid waste (Cooney and Wise, 1975) sponsored by Consolidated, and a detailed engineering analysis of fuel gas production from solid waste (Kispert et al., 1975), funded by the National Science Foundation.

The heart of the anaerobic digestion process is the microbiological conversion of the organic constituents of wastes to methane. The chemistry involved, for any general waste carbohydrate material, may be represented by the following generalized equation:

$$C_nH_aO_b + (4n - a - 2b)/4\ H^2O \rightarrow (4n - a + 2b)/8\ CO_2 + (4n + a - 2b)/8\ CH_4 \qquad (1)$$

Under these circumstances, a pound of waste converted will yield 6.50 cubic feed of methane at standard conditions of temperature and pressure. The methane in this case will be accompanied by an equal volume of carbon dioxide. In general, it has been found that this composition of methane fuel gas is typical for the conversion of a number of organic materials.

MICROBIOLOGY OF ANAEROBIC DIGESTION

Anaerobic digestion has been described as a three-step process in which complex organic materials are converted to the end products of methane and carbon dioxide. A schematic of the process is shown in Figure 1 (Bryant 1976). In general, two groups of microorganisms are responsible for this conversion. The first group of organisms is collectively termed acid formers. These bacteria will convert large organic molecules such as proteins, starches, cellulose, etc., into organic acids (Steps 1 and 2). The organisms performing these steps are both anaerobic and facultative in nature. Step 3, the conversion of acetate to methane and carbon dioxide, is performed by a group of organisms collectively termed methane formers. These organisms are strict anaerobes. Because these methanogens grow more slowly than do the acid formers, Step 3 of this process has been termed the rate limiting step.

As an example, the conversion of a carbohydrate such as glucose (which is formed from the hydrolysis of cellulose or starch) to methane and carbon dioxide is shown by the following equation:

$$C_6H_{12}O_6 \rightarrow 3CH_4 + 3CO_2 \qquad (2)$$

The pathways by which this overall reaction occurs are shown in Figure 2. For every mole of glucose utilized, three moles of methane and three moles of carbon dioxide are formed. For other substrates, the proportions of methane and carbon dioxide produced are different. For example, the conversion of proteins yields a gas mixture with 75 percent methane, while the conversion of fats yields a gas mixture with 70 percent methane (Konstandt, 1976). Thus, the conversion of a complex organic substrate will yield a gas which is typically 50 to 60 percent methane and 40 to 50 percent carbon dioxide.

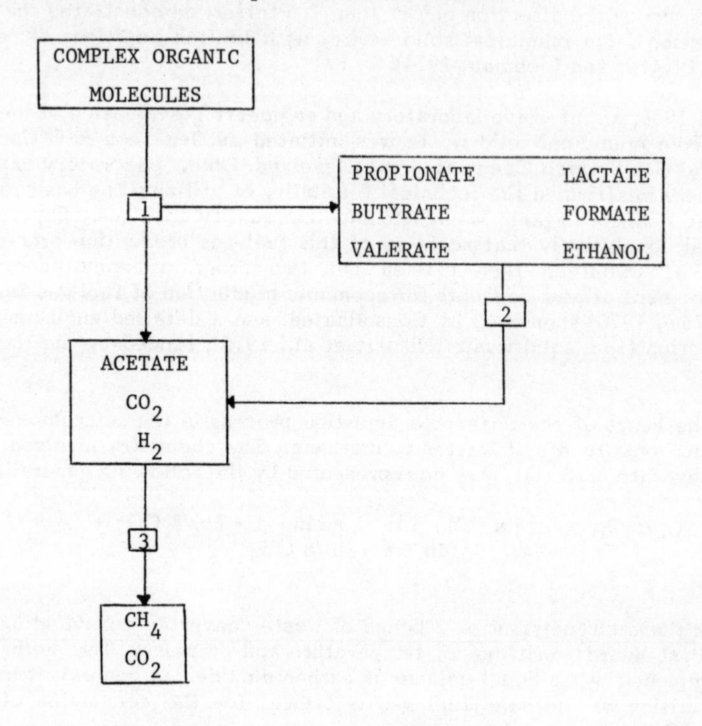

Figure 1: Anaerobic fermentation of complex organics to methane and carbon dioxide (from Bryant, 1976).

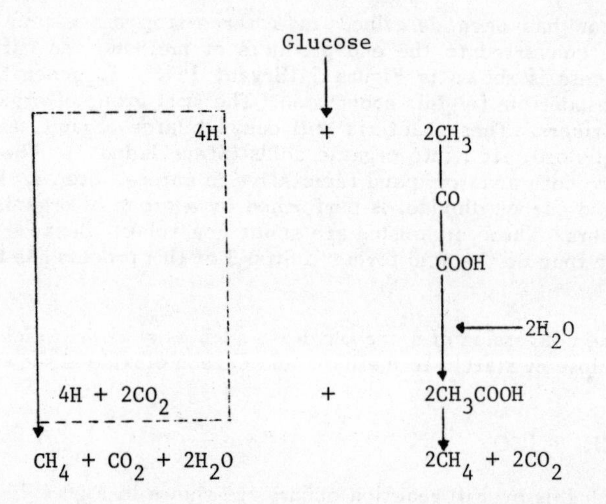

Figure 2: Conversion of glucose to methane and carbon dioxide gas via anaerobic digestion (from Bryant, 1976).

ENVIRONMENTAL CONSTRAINTS

Environmental parameters of concern for providing proper anaerobic digestion conditions include temperature, pH, alkalinity, volatile acid concentrations, nutrients, and presence of toxic substances. Due especially to the sensitivity and slow growth of the methane formers, these environmental parameters must be controlled within specific ranges for adequate digestion to occur.

TEMPERATURE

Three specific temperature ranges exist for anaerobic fermentation, psychrophilic (10 to 20°C), mesophilic (30 to 40°C), and thermophilic (50 to 60°C). In a hole-in-the-ground type digester, the temperature cannot be controlled as it can be in a conventional digester. However, due to the exothermic nature of the anaerobic reactions, these systems will generally exhibit higher temperatures than the surrounding atmosphere. The digestion process may occur more slowly than under optimal conditions (either mesophilic or thermophilic), but it does not cease.

PH, ALKALINITY, AND VOLATILE ACIDS

These three environmental parameters are interdependent. For stable anaerobic digestion to occur the system pH should be between 6.6 and 7.6, with an optimum range of 7.0 to 7.2 (McCarty, 1964). Volatile fatty acids, intermediate products in the fermentation process, can increase in concentration if a system imbalance occurs due to other environmental factors which inhibit the growth of the methane formers. In such a situation, the volatile acids will increase in concentrations and the pH will drop, thus further aggravating the situation until methanogenesis ceases. However, this problem can be avoided if buffer is added to keep the pH near neutral.

NUTRIENTS

A variety of organic and inorganic substances are required for adequate digestion to occur. Among these are carbon, nitrogen, phosphorus, sulfur, vitamins, and trace minerals (Loehr, 1966). Agricultural residues and municipal solid wastes may be deficient in one or more of these nutrients and require supplementation. Of specific concern is the ratio of carbon to nitrogen to phosphorus - (C:N:P); a ratio of 100:5:1 is supposed to be adequate in most cases (Sanders and Bloodgood, 1963; van den Berg and Lentz, 1977; Lane, 1979).

TOXIC SUBSTANCES

Numerous substances, if present in a high enough concentration will cause inhibitory or toxicity problems. Those of major concern include sulfides (>200 mg/L), soluble heavy metals (>1.0 mg/L), alkaline earth metals such as sodium (5,000 - 8,000 mg/L), potassium (4,000 - 10,000 mg/L), calcium (2,000 - 6,000 mg/L), and magnesium (1,200 - 3,500 mg/L), and ammonia (1,700 - 4,000 mg/L) (EPA, 1974).

With agricultural and municipal residues, these substances should not cause significant problems with the possible exception of localized areas within the digester. Most of these substances only become soluble (and therefore toxic) at low pHs, and through adequate buffering the pH should remain near neutral.

PROCESS DESCRIPTION

It has long been known that methane gas is produced in landfills. However, only recently have landfills been viewed as methane producers whose energy is economically recoverable. Subsequently, the art of landfill design has evolved to the point where consideration is being given to designing landfills with energy recovery in mind. Applying landfill gas recovery concepts for energy production from agricultural residues in LDCs hold many significant advantages:

1. Landfill construction is not capital intensive. In the construction of a landfill, the soil is used to contain the residue. This eliminates the need for expensive digesters constructed from steel or concrete. By lining the bottom and top of a landfill with a material such as compacted clay, anaerobic conditions are maintained. In addition, the liner serves as a barrier to liquid migration to the groundwater, and loss of methane gas to the atmosphere.

2. Landfills require no residue pretreatment. Conventional digestion systems, whether CSTR, two stage or plug flow digester all require some form of residue pretreatment. Initially, the residue would have to undergo particle size reduction via grinding or chopping. Secondly, the residue would have to be slurried so that it could be pumped in and out of a digester. Landfilling of the residue requires no pretreatment of this form.

3. Landfills are labor-intensive, low technology systems. The construction of a landfill requires manpower for placement and covering of residues. Once the residue is in place only minimal attention is required to keep the system operating, and no sophisticated technology is needed to maintain gas production, or recover the gas. In comparison, conventional digestion, or the production of other energy forms from agricultural residues (such as alcohols) require a higher degree of technical sophistication.

4. No residue disposal is required. With other energy recovery systems, the residue after fermentation requires some form of liquid/solid separation and solids disposal. In this approach the residue is already in place in a landfill requiring no further disposal.

Therefore, it is the objective of this program to explore the feasibility of utilizing a landfill approach for the production of methane from agricultural residues for LDCs. Dynatech has pioneered the development of various landfilling techniques designed to improve both the ultimate methane yield from organic material and the rate at which methane is produced from organics. The following review of landfill gas recovery serves as a description of how this technology is applied to municipal solid wastes.

LANDFILL GAS RECOVERY

The world's first commercial landfill methane recovery facility started in 1971 when the Los Angeles County Sanitation District constructed wells to prevent gas seepage to adjacent residential properties at the Palos Verdes landfill, in California. Since that time a number of other projects have been initiated. In addition to the Palos Verdes landfill, the ones in commercial operation include the Mountain View landfill (Mountain View, California), the Azusa-Western landfill (Azusa, California), and the Sheldon-Arleta landfill (Sun Valley, California). Other landfills on line include Ascon, California; City of Industry landfill, California; Operating Industries landfill, California; and Cinnamansion landfill, New Jersey (Yoshioka, 1980).

The first step in recovering methane from landfills is to drill down into the landfill and install gas extraction wells. After drilling, the gas extraction wells are typically lined with a perforated casing and packed with sieved gravel. Collection manifolds connect the wells, bringing the "biogas" to a common site. This method has proved to be generally satisfactory.

Removal of gases from existing landfills has been accomplished by the same principle that is involved in the extraction of ground water. Pumps are manifolded to the wells, creating a pressure gradient within the landfill which draws the gas through the collection systems. It is necessary not to draw on the wells at such a rate as to draw air back down into the landfill. Oxygen is detrimental to the anaerobic process and can severely retard methane production.

After extraction, the gas may be upgraded. A typical system consists of dehydration by compression and cooling, pretreatment for hydrogen sulfide and free water removal in molecular sieve towers, followed by another series of molecular sieve towers, followed by another series of molecular sieve absorption towers to remove carbon dioxide. The resulting methane gas (35 GJ/m^3) is pressurized to 2.5 MPa and injected into a pipeline distribution system. Corrosion problems have been encountered in field installations. However, it is expected that corrosion can be eliminated with the injection of inhibitors into the biogas before processing.

As noted earlier, evaluations of the extraction of methane from existing uncontrolled landfills for the Mountain View Project (Blanchet, 1979), the project at Palos Verdes, and by the Los Angeles Bureau of Sanitation (Bowerman et al., 1976) have been made. The low rate of methane production from these landfills results in high costs for the gas. Typical gas production from the existing uncontrolled landfills is in the range 3.8-14.3 m^3/Mg year of deposited waste, with a methane content above 50%. Costs were based on a 20-year life span of a gas recovery system at an existing landfill.

Without enhancement, a landfill undergoes active degradation over a period of many years, the time primarily dependent on the moisture content of the fill. Thus, problems associated with a given landfill (e.g., methane migration, ground water contamination) are likely to continue far beyond the termination of refuse placement, and indeed, may not appear for some time after fill completion. Postconstruction maintenance and monitoring of a completed landfill site, to ensure that it maintains its integrity and does not become a source of pollution, can therefore be a long-term source of concern. Another disadvantage of slow degradation is the long period of time required for gas extraction and the low annual methane yields per mass of refuse. Significant cost savings could be realized by increasing the rate of degradation within existing landfills. Research to date has indicated that higher ultimate gas yields are possible in controlled landfills.

The concept of enhancing methane production in landfills, controlled landfilling as it is sometimes known, has been developed by considering the landfill as a large batch anaerobic digestion system in which optimum conditions for methane production are provided. Urban refuse, which may have been separated, shredded, or baled is combined with nutrients, buffer, and innoculum before its deposition into the landfill for the purpose of sustaining high reproductive rates of bacteria during decomposition. The landfill is constructed to allow for a gas recovery system and a moisture control system, and to optimize refuse cell size and geometry. The gas is extracted once the refuse cell is anaerobic and decomposition begins (DeWalle et al., 1978).

As with the other digestion processes, the composition of the refuse directly affects the rate of methane production, and subsequently, the methane yield. It is advantageous for the refuse to have a high concentration of biodegradable materials, such as food, garden wastes, and paper. Sewage sludge admixed with the refuse increases the fraction of biodegradable materials in the landfill, and at relatively low concentrations (75-400 mg/L), stimulates gas production. At higher concentrations, however, inhibition may result (McCarty, 1964).

The nutrients, buffer, and innoculum may be provided either by chemicals or sewage sludge, which are mixed or layered with the MSW before its deposition into a landfill, or by recycling leachate through the landfill, or by a combination of these methods.

Since the bacteria which carry out the biodegradation process grow best within a narrow pH range, the pH within the landfill should be controlled. The optimal pH range of 6.25 to 7.5 has been controlled by the addition of calcium carbonate, a buffering agent, in simulated landfill cells (Augenstein et al., 1976). Based on a bacterial cell formula of $C_5H_7NO_2$, about 12.4% by weight nitrogen is needed for cell growth, while one-fifth of that value of phosphorus is required (Augenstein et al., 1977). The nutrient value of recycled leachate depends on the landfill composition. The available evidence suggests that the leachate material could provide only a portion of the nutrient requirements; thus, the addition of sewage or artificial nutrients may also be necessary. However, it has been shown by others that macro-nutrient supplementation, i.e., addition of N, P, S, is unnecessary (LeRoux and Wakerly, 1978). The application of sewage sludge, and the recirculation of leachate on controlled landfills is a potential health hazard by spreading bacteria and viruses, and a potential odor nuisance to nearby residents.

The two most important factors affecting methane production rates in landfills are temperature and moisture content. Methane production is severely limited at temperatures below 15°C, but increases with increasing temperature to an optimal temperature of 30-40°C. This parameter cannot, however, be easily controlled in landfills. Several researchers believe that, even in colder climates, significant temperatures can be reached due to thermal insulation from surrounding soil and refuse (Rees, 1980). High temperature areas are expected to be non-homogeneous in distribution.

The refuse moisture content should be at least 50% and preferably about 80%, for high methane yields (Augenstein et al., 1976). Methane production increases exponentially with increases in moisture content in batch digestion (DeWalle et a., 1978). The moisture content of refuse (normally 25%) may be increased by the addition of water, sewage sludge, industrial wastes, or leachate material before deposition into the landfill. Alternatively, the natural ground water supply could provide water during digestion, but this may contaminate the water as discussed above. Although methods for water addition have been evaluated, large scale implementation of moisture control systems is needed.

Whatever the source of moisture, the landfill must be designed to retain moisture and to obstruct the flow of polluting leachate material. Impermeable or low-permeable barriers, such as certain types of soils or synthetic liners, can be placed on the bottom and sides of the landfill area; these barriers are commercially available. The landfill should slope towards a point where leachate material can be collected for treatment or collection. If the topography of the landfill site does not provide for natural drainage, recontouring the land, probably at great expense, is necessary.

The enhancement concept appears to be a potentially effective method for producing methane from municipal solid waste. Experiments using laboratory test cells have provided information on the effects of controlled conditions on methane production. Field scale demonstration facilities employing this concept are presently in operation at the Mountain View landfill in California and at the Binghampton, New York landfill.

APPLICATION OF ENHANCEMENT TO AGRICULTURAL RESIDUES

In a recently completed study at Cornell University (Jewell et al., 1981), the principles of microbial innoculation, buffer additions, nutrient addition, and moisture requirements were assessed for digestion of wheat straw and corn stover. The effect of these parameters on the anaerobic degradation of these agricultural residues were assessed in reactors similar to a landfill-type environment. It was the attempt of this study to show that, digestion of agricultural residues could be achieved under minimal moisture conditions (i.e., not having the residues in a slurry) under appropriate conditions. The need for adequate microbial innoculum and buffer additions to achieve rapid start-up of digestion and efficient conversion of these residues, was found to be crucial in this study.

REFERENCES

Augenstein, D.C., Wise, D.L., Wentworth, R.L. and Cooney, C.L. 1976. Fuel Gas Recovery From Controlled Landfilling of Municipal Wastes. Resource Recovery and Conservation, 2: 103-117.

Augenstein, D.C., Wise, D.L., Wentworth, R.L., Gallaher, P.M. and Lipp, D.C. 1977. Investigation of converting the product of coal gasification to methane by the action of microorganisms. Final Report no. FE-2203-17 under contract no. E.(49-18)-220. Dynatech R/D Company, Cambridge, MA.

Blanchet, M. 1979. Start-up and operation of the landfill gas treatment plant at Mountain View. Solid and Hazardous Waste Research Division 5th Annual Research Symp., Municipal Solid Waste: Resource Recovery, Orlando, FL.

Bowerman, F.R., Rohatji, N.K. and Chen, K.Y., (Eds.) 1976. A case study of the Los Angeles Sanitation Districts Palos Verdes Landfill Gas Development Project. U.S. Environmental Project. U.S. Environmental Protection Agency, NERC Contract no. 68-03-2143.

Bryant, M.P. 1976. The Microbiology of Anaerobic Digestion and Methanogenesis with Special Reference to Sewage. In: Seminar on Microbial Energy Conversion, Gottingen, Germany.

Buivid, M.G., Wise, D.L., Blanchet, M.J., Remedios, E.C., Jenkins, B.M., Boyd, W.F., Pacey, J.G. 1981. "Landfillings of Municipal Solid Waste", 6: 3-20.

Cooney, C.L., and Wise, D.L. 1975, Bioengineering and Biotechnology 17:1119.

DeWalle, F.B., Chian, E.S.K. and Hammerberg, E. 1978. Gas Production from solid waste in landfills. J. Envir. Eng. Div. 104:415-432.

EPA 1974. Process Design Manual for Sludge Treatment and Disposal. EPA625/1-74-006, Technology Transfer, Washington, D.C.

Golueke, C.G. 1971. Comprehensive Studies of Solid Waste Management. Third Annual Report, U.S. Environmental Protection Agency, Washington, D.C.

Golueke, C.G. and McGauhen, P.H. 1970, Comprehensive Studies of Solid Waste Management. First and Second Annual Reports, Bureau of Solid Waste Management, Washington, D.C.

Jewell, W.J., Chandler, J.A., Dell'Orto, S., Kanfoni, K.J., Fast, S., Jackson, D. and Kabrick, K.M. 1981. "Dry Fermentation of Agricultural Residues", SERI Report No. XB-0-9038-H6, April.

Kispert, R.G., Sadek, S.E., and Wise, D.L. 1975. An Economic Analysis of Fuel Gas Production. In: Seminar on Microbial Energy Conversion, Gottingen, Germany.

Konstandt, H.G. 1976. Engineering, Operation, and Economics of Methane Gas Production. In: Seminar on Microbial Energy Conversion, Gottingen, Germany.

Lane, A.G. 1979. Methane from Anaerobic Digestion of Fruit and Vegetable Processing Wastes. Food Technology in Australia, 31, 5:201-207.

LeRoux, N. and Wakerly, D. 1978. The Microbial Production of CH_4 From the Putresibile Fractions of Solid Household Waste. Proc. 1st Recycling World Congress (M.E. Henstock, ed.), Basel, Switzerland.

Loehr, R.C. 1966. Design of Anaerobic Digestion Systems. Journal San. Eng. Div. ASCE, (SAI) 92:19-29.

McCarty, P.L. 1964. Anaerobic Waste Treatment Fundamentals: I. Chemistry and Microbiology; II. Environmental Requirements and Control; III. Toxic Materials and Their Control; IV. Process Design. Public Works, 95:9-12.

Pfeffer, J.T. 1974. Reclamation of Energy from Organic Waste. Final Report, U.S. Environmental Protection Agency, Washington, D.C.

Pfeffer, J.T., and Liebman, J.C. 1974. Biological Conversion of Organic Refuse to Methane. Annual Progress Report to National Science Foundation, NSF Grant GI-39191. Washington, D.C.

Rees, J.F. 1980. Journal Chemical Technol. Biotechnol., 30:458-465.

Saunders, F.A., and Bloodgood, D.E. 1965. The Effect of Nitrogen-to-Carbon Ratio on Anaerobic Decomposition. JWPCF, 37 12:1741-1752.

Van den Berg, L. and Lentz, C.P. 1977. Methane Production During Treatment of Food Plant Wastes by Anaerobic Digestion. In: Cornell Agricultural Waste Management Conference, Food, Fertilizer Residues, (R.C. Loehr, ed.) Ann Arbor: Ann Arbor Science Publishers, Inc.

Wise, D.L., Boyd, W.D., Blanchet, M.J., Remedios, E.C., Jenkins, B.M. 1981. "In Situ Methane Fermentation of Combined Agricultural Residues," 6:275-294.

Yoshioka, G. 1980. Paper presented at Landfill Methane Utilization Seminar, Ailomar, Pacific Grove, CA., March.

BUTANOL AND BUTANEDIOL PRODUCTION FROM PRETREATED BIOMASS

Ernest K.C. Yu and John N. Saddler
Biotechnology Department, Forintek Canada Corporation
800 Montreal Road, Ottawa, Ontario K1G 3Z5

ABSTRACT

The utilization of the cellulose and hemicellulose of pretreated biomass for the production of fuels and chemicals was investigated. Aspenwood was pretreated by steam-explosion and then fractionated by water-extraction into a water-extract fraction rich in sugars from hemicellulose and a cellulose-rich residue. The two fractions were then hydrolyzed either by acid or fungal enzymes (cellulase and xylanase enzyme complexes of *Trichoderma harzianum* E58) to component sugars. The sugar mixtures of both the acid and enzyme hydrolyzates of the cellulose or hemicellulose fractions could be fermented efficiently by *Klebsiella pneumoniae* ATCC 8724 and *Clostridium acetobutylicum* ATCC 824 for the production of butanediol and butanol, respectively. The production of butanediol was further enhanced by combining enzymatic hydrolysis and fermentation (CHF). The process was readily scaled up in laboratory fermentors. The process was also successfully extended to the utilization of the combined cellulose and hemicellulose components of aspenwood and agricultural residues (i.e., steam-exploded biomass without further extraction). The results demonstrated the industrial potential of the two organisms for the bioconversion of total biomass carbohydrates to valuable fuels and chemicals.

INTRODUCTION

The full utilization of both the cellulose and hemicellulose carbohydrates of wood and agricultural residues must be achieved if the bioconversion of biomass to fuels and chemicals is to be competitive economically with the petrochemical industry. Moreover, while cellulose is a homogenous polymer of glucose sub-units, hemicelluloses are composed of pentose and hexose sugars, uronic acids, acetyl groups, etc., in various proportions, depending on the source of biomass. It is therefore crucial for an efficient conversion process to select fermentative organisms capable of utilizing all the available substrates for the production of valuable fuels and chemicals. We have recently shown the potential of *Klebsiella pneumoniae* and *Clostridium acetobutylicum* for converting biomass substrates to butanediol, ethanol, acetone, and butanol. This paper reviews and updates the work on butanediol and butanol production from biomass substrates carried out in our laboratory over the past three years.

RESULTS AND DISCUSSION

The major sugars obtained from acid or enzymatic hydrolysis of the hemicellulose and cellulose fractions were shown to be a mixture of pentoses (L-arabinose and D-xylose), hexoses (D-glucose, D-galactose, and D-mannose), as well as low molecular weight oligosaccharides (e.g. cellobiose and xylobiose) (Saddler et al., 1983; Yu et al., 1984c). *K. pneumoniae* could utilize all such sugars for the production of 2,3-butanediol and ethanol

(Yu and Saddler, 1982a). The fermentation was particularly efficient under finite air conditions when acetic acid was added to the culture media (Yu and Saddler, 1982b; 1984c) (Table 1).

Table 1. Butanediol production by K. pneumoniae grown on sugars present in biomass hydrolyzates.

Substrate (50 g/L)	Substrates used (%)	Acetic acid used (g/L)	Butanediol produced (g/L)
D-glucose	100.0	4.9	25.4
D-xylose	98.9	4.9	27.1
D-galactose	96.2	4.9	25.4
D-mannose	98.0	5.0	25.8
L-arabinose	96.8	4.7	26.0
D-cellobiose	96.0	4.7	18.0

The added acetic acid (5 g/L) resulted in enhanced butanediol yields by the combined mechanisms of induction, activation, and inhibition on the three enzymes known to be involved in butanediol metabolism. Moreover, over 95% of the added acetic acid was actually consumed, presumably by its condensation with pyruvate, for the additional production of butanediol. The presence of acetic acid also reduced acid formation during the fermentation process, thereby eliminating the need for additional buffer in the medium by MES (2-[N-morpholino]-ethanesulfonic acid). This resulted in a simpler and much cheaper medium for large scale production of butanediol.

One major disadvantage in butanediol production on an industrial scale has been the high cost of its product recovery as a result of the high boiling point of butanediol (around 180 C). Attempts were therefore made to maximize the final product levels of the bioconversion process. By adopting a double fed-batch approach (daily addition of sugar and yeast extract) using K. pneumoniae cells previously acclimatized to high initial substrate concentrations, butanediol levels of 106 g/L and 80 g/L were obtained from the consumption of 226 g/L of D-glucose and 190 g/L of D-xylose respectively (Yu and Saddler, 1983a). A further improvement in the economics of the process was obtained by replacing the relatively costly yeast extract by the inorganic nitrogen source ammonium sulfate (Fig. 1). The final butanediol concentrations under these conditions were 114.7 g/L and 74.4 g/L from D-glucose and D-xylose, respectively. This surpassed the levels which have been estimated to be required for economic product recovery. Comparable product yields were obtained when laboratory fermentors were used.

We had previously shown that the sugars present in biomass hydrolyzates could be converted by C. acetobutylicum to butanol, acetone, and ethanol (Mes-Hartree and Saddler, 1982; Yu and Saddler, 1983b). By including a low level of acetic acid (1 g/L) in the fermentation media, butanol yields of 13.7 g/L could be obtained from 60 g/L of D-glucose, corresponding to 0.26 g of butanol per g of sugar consumed (Table 2). D-xylose utilization was less efficient, as reflected by the lower butanol concentrations.

These encouraging results prompted us to study the possibility that K. pneumoniae and C. acetobutylicum could utilize both the cellulose and hemicellulose components of the lignocellulosic residues. Previously (Saddler et al., 1982a) we had shown that steam-exploded aspenwood could be separated into water extracts rich in sugars from hemicellulose (SEW-WS) and cellulose-rich residues (SEW-WI). Both fractions were then subjected to hydrolysis by sulfuric acid and the sugars derived were fermented by C. acetobutylicum and

K. pneumoniae (Yu et al., 1984e). Water extracts of steam-exploded aspenwood, aspenwood holocellulose and aspenwood were hydrolyzed by 3% (w/v) sulfuric acid, and the liberated sugars were used as the substrate of growth for C. acetobutylicum. Butanol yields were only 40-60% of the theoretical (Table 3), presumably because of the relatively high proportion of pentose sugars that were likely to be present in these hydrolyzates. Higher butanol yields were obtained when cellulose (water-insoluble residues of steam-exploded aspenwood) or aspenwood was directly hydrolyzed by concentrated sulfuric acid (Yu et al., 1984d).

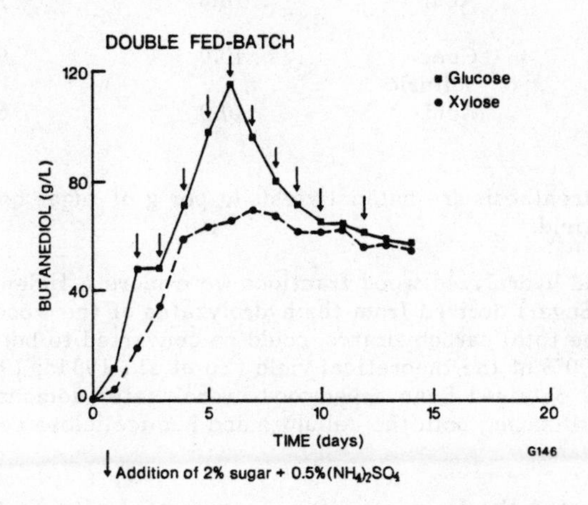

Figure 1. Double fed-batch approach to butanediol production in the absence of MES and yeast extract.

Table 2. Solvent production by C. acetobutylicum grown on sugars present in biomass hydrolyzates.

Substrate (40 g/L)	Butanol (g/L)	Butanol (g/g)**	Acetone (g/L)	Ethanol (g/L)
D-glucose *	13.7	0.26	4.5	2.2
D-xylose *	6.0	0.22	2.1	0.9
L-arabinose	7.1	0.24	3.4	0.9
D-mannose	10.2	0.27	3.3	1.1
D-cellobiose	9.0	0.25	2.9	0.9

(All solvent columns under "Solvents Produced".)

* Cultures grown with added acetic acid
** Butanol yields (g of butanol produced per g of sugar utilized). Theoretical butanol yield calculated as 0.27.

Table 3. Butanol production from acid hydrolyzates of aspenwood.

Substrate	Hydrolysis Conditions	Sugar Conc'n (g/L)	Butanol produced (g/L)
SEW-WS	Dilute	10.0	0.8 (40.7)*
Holocellulose	Sulfuric	10.0	3.0 (59.3)
Aspenwood	Acid	10.0	2.8 (55.6)
SEW-WI	Conc.	40.0	9.0 (92.6)
	Sulfuric		
Aspenwood	Acid	40.0	6.6 (78.9)

* Values in parenthesis are butanol yields (g per g of sugar consumed) as percentages of the theoretical yield.

The various acid hydrolyzed wood fractions were more efficiently utilized by K. pneumoniae (Table 4). Sugars derived from the hydrolyzates of the wood hemicellulose and cellulose, as well as the total carbohydrates, could be converted to butanediol resulting in values approximately 90% of the theoretical yield (Yu et al., 1984d; 1984e). The final product level of 5.1% (w/v) obtained from aspenwood hydrolyzates demonstrated the potential of this organism for utilizating both the cellulose and hemicellulose components of lignocellulosic substrates.

Recently we tested the fermentability of aspenwood after it had been hydrolyzed by anhydrous hydrogen fluoride (HF). Hydrolyzates from HF-treated aspenwood were predominantly composed of oligosaccharides which could not be directly utilized by K. pneumoniae (Yu et al., 1984a). Attempts at further hydrolyzing these oligosaccharides using a variety of glycolytic and xylanolytic enzymes (i.e., amylases, cellulases, and xylanases) were only partially effective. However, when a simple post-hydrolysis was carried out using dilute sulfuric acid, over 85% (w/w) of the HF-treated substrate was released in the form of monosaccharides and these sugars were readily fermented by the organism (Yu et al., 1984a) (Table 4). The high recovery of oligosaccharides after the HF-treatment, the high yield of the post hydrolysis step, and the resulting fermentability of the sugars in the hydrolyzates indicated that this approach may have some potential as a future process for converting wood to fuels and chemicals.

Table 4. Butanediol production from acid hydrolyzates of aspenwood.

Substrate	Hydrolysis Conditions	Sugar Conc'n (g/L)	Butanediol produced (g/L)
Xylan	Dilute	20.0	8.2 (90.0)*
SEW-WS	Sulfuric	20.0	9.9 (92.0)
Aspenwood	Acid	20.0	6.5 (72.0)
SEW-WI	Conc.	100.0	46.0 (94.9)
	Sulfuric		
Aspenwood	Acid	125.0	51.2 (88.4)
Aspenwood	HF	50.0	7.6 (45.8)

* Values in parenthesis are butanediol yields (g per of sugar utilized) as percentages of the theoretical yield.

We have also studied enzymatic hydrolysis as an alternative means of obtaining fermentable sugars from the cellulose and hemicellulose polysaccharides. Earlier we had shown that Trichoderma harzianum E58 released a full spectrum of cellulase and xylanase enzymes (Saddler, 1982) which could efficiently hydrolyze cellulose and hemicellulose from a variety of sources (Saddler et al., 1983a; 1983b). However, the hydrolysis efficiencies declined with increasing substrate concentration (Saddler et al., 1983b). This was presumably a result of the accumulation of hydrolysis end-products (i.e., glucose, cellobiose, xylose, and xylobiose) which inhibited further enzymatic hydrolysis. Once the substrates were hydrolyzed, the sugars derived from wood hemicellulose and cellulose could be fermented by both K. pneumoniae and acetobutylicum (Saddler et al., 1983a; 1983b; Yu et al., 1984e). This indicated that the overall bioconversion efficiencies were limited more by substrate hydrolysis than by the subsequent fermentation. Recent research on ethanol production from wood cellulose in our laboratory (Saddler et al., 1982b; Mes-Hartree et al., 1983) showed that the enzymatic hydrolysis efficiency could be improved by adopting a combined hydrolysis and fermentation (CHF) approach. A similar technique was used with hemicellulose conversion to butanediol (Yu et al., 1984b; 1984c) and found to enhance solvent production significantly (Table 5). This approach was also successful for butanediol production from wood cellulose (Yu et al., 1984c), using solka floc as the model substrate. Preliminary studies using laboratory fermentors also showed that the CHF process could be easily scaled up.

Table 5. Butanediol production from xylan and solka floc using the sequential hydrolysis and fermentation (SHF) or the combined hydrolysis and fermentation (CHF) approaches.

Substrate	Substrate Conc'n (g/L)	Butanediol produced	
		SHF	CHF
Xylan	50	5.7	8.2
	100	9.9	16.6 (18.5)*
Solka floc	50	6.4	9.1
	100	10.4	15.0 (22.6)*

* Values in parenthesis were obtained from fermentor studies using the CHF approach.

The CHF approach could also be used for the conversion of hemicellulose and cellulose to butanol. Aspenwood xyland could be hydrolyzed by xylanase(s) present in the fungal culture filtrates using incubation conditions which had been previously optimized for butanol production (Table 6) and the reducing sugars released were readily fermented to butanol, acetone, and ethanol. Butanol yields from xylan were generally low, again reflecting the lower efficiency of xylose conversion. However, under the same conditions, solka floc could be readily fermented to butanol by the CHF process. These results indicated the potential of using this process for the combined utilization of the hemicellulose and cellulose components of biomass substrates.

Table 6. Butanol production from xylan and solka floc using a combined hydrolysis and fermentation approach.

Substrate (50 g/L)	Reducing sugars present (g/L)		Solvents produced (g/L)		
	Hydrolysis	CHF	Butanol	Acetone	Ethanol
Xylan	30.6	14.2	3.4	1.0	0.6
Solka floc	25.1	1.6	6.5	3.1	0.7

The feasibility of using the CHF approach for the production of butanediol from wood hemicellulose, cellulose, and ultimately the combined cellulose and hemicellulose carbohydrates, was tested (Yu et al., 1984b; 1984c). High butanediol yields (g of product per g of original untreated biomass substrate) were obtained from the steam-exploded aspenwood and agricultural residues (Table 7).

Table 7. Solvent production from steam-exploded substrates by the CHF approach.

Steam-exploded Substrate	Yields (g/g of original substrate)	% Theoretical conversion
Aspenwood	0.23	65.4
Wheat straw	0.20	64.2
Barley straw	0.14	38.9
Corn stover	0.12	42.9

Recent studies also showed that the product yields obtained from the unextracted steam-exploded substrates surpassed the combined yields obtained from the hemicellulose sugars and the water-insoluble cellulose fractions (Yu et al., 1984c). At higher initial substrate concentrations, however, product yields from the unextracted steam-treated substrates decreased. The inhibition was subsequently relieved by water-extracting the steam-treated substrates, suggesting the presence of water-soluble inhibitors formed during the steam-explosion treatment. This was confirmed when the utilization of D-glucose and D-xylose by the organism was inhibited in culture media supplemented with varying quantities of the water-extracts from steam-exploded substrates (Saddler et al., 1983a). The identities of the inhibitors have not yet been determined. Recent studies in our laboratory showed that furfural and hydroxymethylfurfural (HMF) both resulted in decreased cell growth, sugar utilization, and butanediol production. Under the established fermentation conditions, 50% inhibition in butanediol production resulted from the addition of 2.0 g/L of furfural and 10.0 g/L of HMF on D-glucose grown cells, and 0.5 g/L of furfural and 5.0 g/L of HMF on D-xylose grown cells. These results suggested that furfural and HMF were not solely responsible for the observed inhibition since these furan derivatives are normally detected at concentrations of <0.20 g/L and <0.05 g/L, respectively, in the water-extracts of steam-exploded lignocellulosic substrates (at concentrations of 5%, w/v). Similar findings were also obtained from studies on the effect of furan on acetone-butanol fermentation. We are currently trying to establish a process development unit (PDU) which will allow us to evaluate the technical and economic feasibility of integrating the various steps outlined in Figure 2.

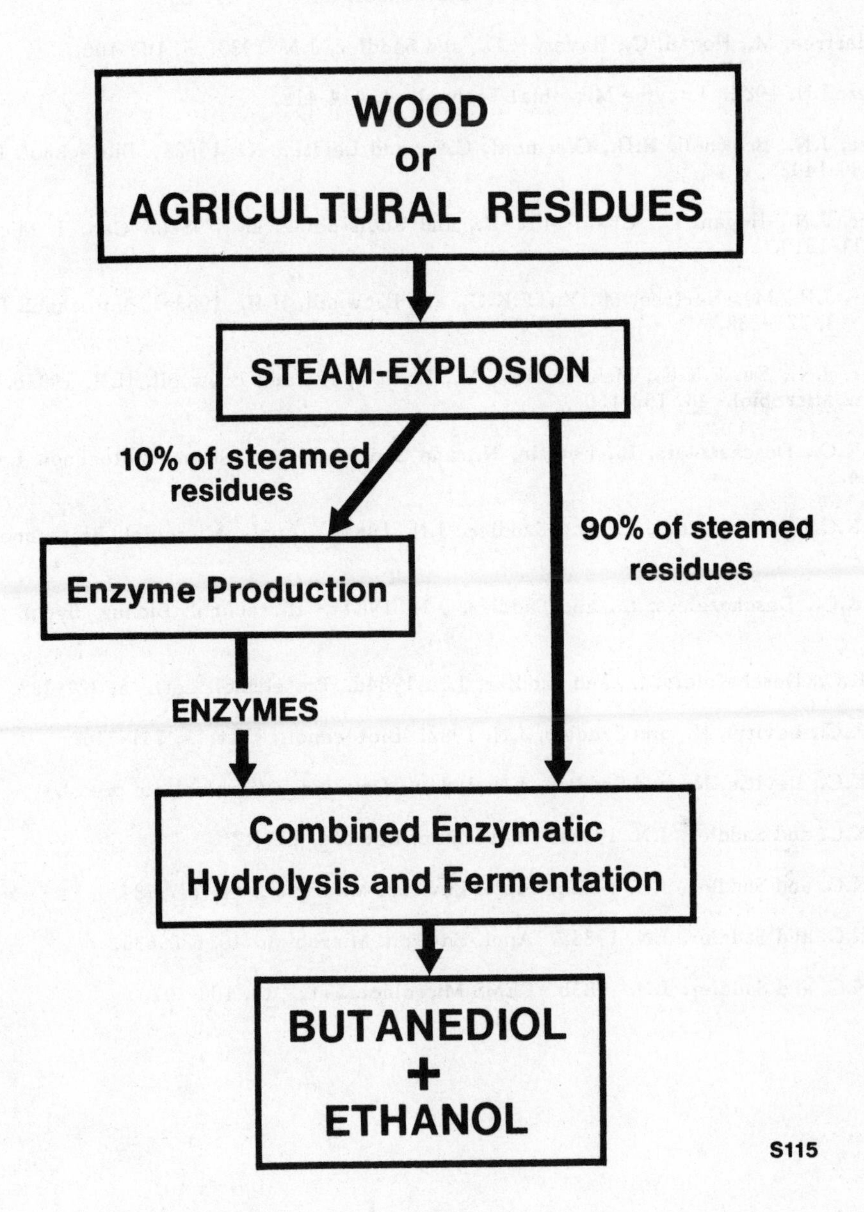

Figure 2. Proposed Process Scheme for the Production of Butanediol from Lignocellulosic Substrates

REFERENCES

Mes-Hartree, M. and Saddler, J.N. 1982. Biotechnol. Lett. 4, 247-252.

Mes-Hartree, M., Hogan, C., Hayes, R.D., and Saddler, J.N. 1983. 5, 101-106.

Saddler, J.N. 1982. Enzyme Microbial Technol. 4, 414-418.

Saddler, J.N., Brownell, H.H., Clermont, C.P., and Levitin, N. 1982a. Biotechnol. Bioeng. 24, 1389-1402.

Saddler, J.N., Hogan, C., Chan, M.K.-H., and Louis-Seize, G. 1982b. Can. J. Microbiol. 28, 1311-1319.

Saddler, J.N., Mes-Hartree, M., Yu, E.K.C., and Brownell, H.H. 1983a. Biotechnol. Bioeng. Symp. 13, 225-238.

Saddler, J.N., Yu, E.K.C., Mes-Hartree, M., Levitin, N., and Brownell, H.H. 1983b. Appl. Environ. Microbiol. 45, 153-160.

Yu, E.K.C., Deschatelets, L., Levitin, N., and Saddler, J.N. 1984a. Biotechnol. Lett. 6, 611-614.

Yu, E.K.C., Deschatelets, L., and Saddler, J.N. 1984b. Appl. Microbiol. Biotechnol. 19, 365-372.

Yu, E.K.C., Deschatelets, L., and Saddler, J.N. 1984c. Biotechnol. Bioeng. Symp. 14 (in press).

Yu, E.K.C., Deschatelets, L., and Saddler, J.N. 1984d. Biotechnol. Lett. 6, 327-332.

Yu, E.K.C., Levitin, N., and Saddler, J.N. 1982. Biotechnol. Lett. 4, 741-746.

Yu, E.K.C., Levitin, N., and Saddler, J.N. 1984e. Dev. Ind. Microbiol. (in press).

Yu, E.K.C. and Saddler, J.N. 1982a. Biotechnol. Lett. 4, 121-126.

Yu, E.K.C. and Saddler, J.N. 1982b. Appl. Environ. Microbiol. 44, 777-784.

Yu, E.K.C. and Saddler, J.N. 1983a. Appl. Environ. Microbiol. 46, 630-635.

Yu, E.K.C. and Saddler, J.N. 1983b. FEMS Microbiol. Lett. 18, 103-107.

Section 3

Production of SCP

PRODUCTION OF FOODS, FOOD ADDITIVES AND FEEDS FROM BIOMASS BY MICROBIOLOGICAL PROCESSES

John H. Litchfield

Battelle-Columbus Laboratories
Columbus, Ohio 43201 U.S.A.

INTRODUCTION

Why did some large scale processes for single-cell protein (SCP) production for food or feed applications fail to survive? Availability and cost of n-alkane substrates was a significant factor in the commercial failure of the British Petroleum and the Kanegafugi Chemical Industry Co. processes for large scale production of *Candida* spp. for use in animal feeds (Litchfield, 1980, 1983a, b). Regulatory problems associated with hydrocarbon residues in the final microbial protein products were also a problem in these processes. In contrast, biomass resources including agricultural, food processing waste and forestry products and their wastes are attractive low-cost sources of carbohydrate raw materials for producing foods, food additives and feeds by microbiological conversion processes.

This discussion covers developments in the production of food and feed-grade microbial protein products (single cell proteins) and selected food additives from biomass-derived carbohydrates by nonphotosynthetic microorganisms including actinomycetes, bacteria, molds, yeasts, and higher fungi. Algal biomass production is discussed in a paper by Burrell et al. (1984) in this symposium.

KEY CONSIDERATIONS IN FOOD, FOOD ADDITIVE, AND FEED PRODUCTION BY MICROBIOLOGICAL PROCESSES

Figure 1 shows some of the most important considerations in producing foods, food additives, and feeds by microbiological processes. Each of these considerations has an important bearing on the ultimate economic viability of microbiological processes for producing food and feed-related products.

Raw Materials

Using plentiful, cheap raw materials is a key factor in the success of microbial protein production processes since raw material costs may be as much as 50 percent or more of the operating costs for the process. Table 1 lists some of the common carbohydrate raw materials that have been considered for use in microbial protein processes. Simple sugars can be utilized by a wide range of microorganisms for growth. However, starchy and lignocellulosic raw materials must be treated by acid or enzyme hydrolysis, alkali pretreatment, or physical pretreatment, or a combination of these methods to yield fermentable sugars

(Bungay, 1983; Ladish et al., 1983). Cellulose and hemi-cellulose must be separated from lignin before hydrolysis to simple sugars.

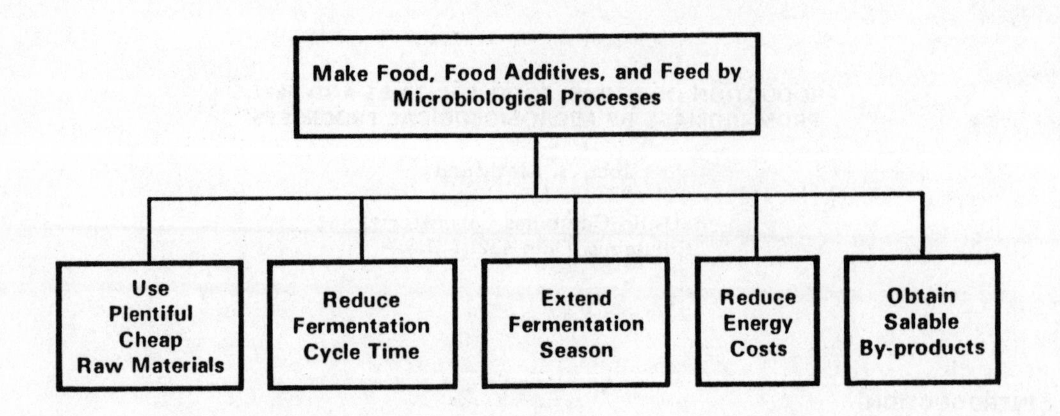

Figure 1: Important considerations in producing food, food additives and food by microbiological processes.

TABLE 1. RAW MATERIALS FOR MICROBIAL PROTEIN PRODUCTION FROM BIOMASS

Raw materials
Sugars
Cane, or beet sugar and molasses, cheese whey (lactose), fruit processing wastes, sulfite water liquor
Starches
Potato, cassava, corn (maize), wheat, vegetable processing wastes
Cellulose and Hemicellulose
Vegetable processing and agricultural wastes (wheat straw, rice straw), forestry and wood wastes

Figure 2 shows some of the products that can be derived from cellulose or hemicellulose. Cellulose can be assimilated directly by cellulolytic microorganisms or, after conversion to glucose, can be used as a substrate for producing SCP or other microbial products of interest as food additives. Hemicellulose can be utilized directly for SCP production by microorganisms that can hydrolyze this substrate or after conversion to pentoses, for the production of SCP or other microbial products.

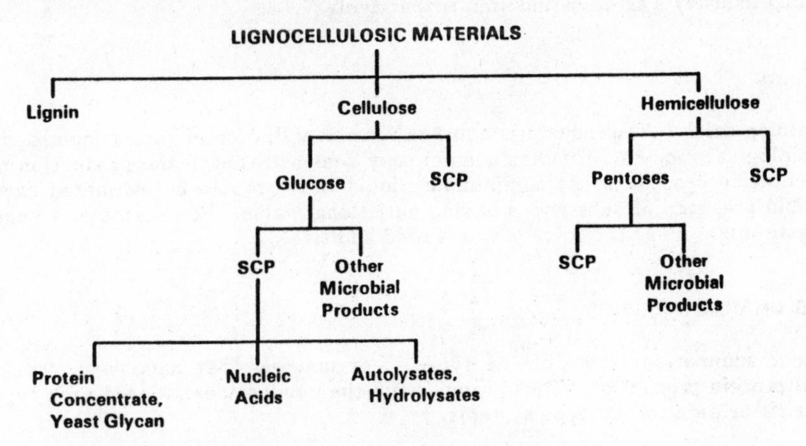

Figure 2: Products of microbiological conversion of lignocellulosic materials.

Some of the desirable characteristics of substrates for microbial production of foods or food additives include the following:

(1) availability - the substrate must be readily available in a uniform supply over the entire year.
(2) composition and physical characteristics - the substrate should have low concentrations of furfural, hydroxymethylfurfural, ash and heavy metal contents. Also, the substrate should be readily handled by commercial bulk handling equipment.
(3) performance - the substrate should give commercially attractive growth rates, and yields with the microorganism of choice.
(4) cost - there is a trade-off among cost of substrate, yield of product, and the selling price of the product that must be considered in selecting the substrate for a particular process.

Fermentation Cycle Time

The cost of the fermentation vessel represents a significant portion of the capital cost of a microbiological process. Consequently, it is important that fermenters be utilized to the maximum extent possible during an operating year. Down time of the fermenter must be reduced as much as possible. A total of 8,000 hours (333 days) of operation is a desirable goal that is not often achieved.

Fermentation Season

Another desirable goal for a microbiological process is to extend the fermentation season over a maximum portion of the year. This means that substrates that are only seasonally available must be avoided unless they can be stored in a stable form in sufficient quantity that the fermentation process can be operated throughout the year. For example, sweet sorghum could be an attractive source of fermentable sugar if methods can be developed for either storing the stalks or the sugar derived from the stalks over a sufficient portion of the year so that the plant can be kept in operation.

Energy Cost

The energy required for SCP production is a significant factor in operating costs. For example, Lewis (1976) calculated the total gross energy requirements in GJ/kg of protein for various processes as follows: Aspergillus niger with solid agricultural waste or agricultural process effluents as substrates, 80 and 30, respectively; and Candida sp. on molasses or distilling industry waste, 75 and 155, respectively.

By-Products

Obtaining salable by-products is another important factor in the economic viability of a microbiological process. This has been clearly demonstrated in the production of ethanol by fermentation processes. At a minimum, liquid waste should be evaluated carefully for the possible presence of substances having nutritional value. For example, a concentrated liquid waste might be suitable for use as a feed additive.

PROCESS CHARACTERISTICS

Table 2 summarizes some of the types of fermenters that have been considered for microbial protein production. These range from the conventional stirred tank-type reactor to the airlift or aeration jet type systems.

TABLE 2. TYPES OF FERMENTERS FOR MICROBIAL PRODUCTION FROM BIOMASS

Stirred tank Aerated, non-agitated	Bakers yeast, primary grown yeast (Saccharomyces cerevisiae) (molasses)
Baffled, agitated, aerated	Torula yeast (Candida utilis), (sulfite waste liquor, ethanol)
Baffled agitated, non-aerated	Lactic acid starter cultures
Airlift (Draft tube)	Torula yeast (Candida utilis)
Aeration jet	Fodder yeast

Table 3 presents characteristics of some selected single-cell protein processes based on carbohydrates. Processes range from batch to continuous. In most batch processes only partial utilization of the carbohydrate feedstock is achieved. In fed-batch or continuous processes the rate of feed of the carbon and energy source can be adjusted to allow complete utilization. The pH is maintained in the acid range, for example at pH 4.5, to minimize contamination problems since these processes are operated in a clean but nonsterile mode.

TABLE 3. CHARACTERISTICS OF SELECTED MICROBIAL PROTEIN PROCESSES BASED ON CARBOHYDRATES[a]

	Process			
	Sulfite Waste Liquor		Cheese Whey	Cane Molasses
Characteristic	*C. utilis*	*P. varioti*	*K. fragilis*	*S. cerevisiae*
Type of process	Continuous or batch	Continuous	Batch	Batch or fed batch
Sterility	Nonaseptic	Nonaseptic	Nonaseptic	Nonaseptic
Fermenter	Agitated or airlift	Agitated	Agitated	Agitated (by aeration
Feedstock utilization	Partial	Partial	Partial	Partial to complete
Temperature (°C)	32-38	38-39	32	30-36
pH	4.5-5.5	4.5-4.7	4.5-5.5	4.5-5.0
Product recovery	Centrifugation	Filtration	Centrifugation	Centrifugation

(a) Litchfield, 1983b

Yeast products are recovered by centrifugation which is relatively costly. Mold and fungal processes have the advantage that the mycelium can be recovered by simple filtration, which is less costly than centrifugation. Recently, Shay and Wegnor (1984) of Phillips Petroleum Co. described a continuous process for producing C. utilis from sugars in which a cell concentration of 150 g/L (dry weight) is achieved with productivity in the range of 30 g/L•hr. The yeast can be recovered by drying without centrifugation. Details have not been published on their novel fermenter design that provides for oxygen transfer rates of over 800 m mol/L•hr.

PRODUCT QUALITY

Food grade microbial protein products include the dried cells of yeast, molds, and higher fungi. Tale 4 presents a list of these products. The yeast products are approved by the Food and Drug Administration for sale as food grade products in the United States. The Fusarium graminearum product developed by Rank Hovis MacDougall has been approved for test market studies in the United Kingdom (Solomons, 1983). Dried morel mushroom mycelium (Morchella sp.) has been approved for sale as a flavouring additive in the USA but is not currently produced owing to economic factors (Litchfield, 1979).

TABLE 4. FOOD GRADE MICROBIAL PROTEIN PRODUCTS

Product	Organism	Substrate	Reference
Yeast	Candida utilis	Sulfite waste liquor, ethanol	Peppler, 1979
	Kluyeromyces fragilis	Cheese whey	Bernstein et al., 1977
	Saccharomyces cerevisiae	Molasses (cane, beet)	Peppler, 1979
Molds and higher fungi	Fusarium graminearum	Glucose syrup	Solomons, 1983
	Morchella sp.	Glucose, cheese whey, cannery wastes	Litchfield, 1979

Table 5 summarizes some processes for making feed grade products from biomass substrates. Both slurry and solid state fermentation have been investigated. None of these processes is practiced on a commercial scale although processes for utilizing bagasse and agricultural, and pulp and paper wastes have been operated at a pilot plant scale.

TABLE 5. FEED GRADE PRODUCTS FROM BIOMASS SUBSTRATES

Substrate	Organism	Type of Process	Yield	Reference
Barley straw	Cellulomonas sp. Candida utilis	Continuous	32-61%[a]	Kristensen, 1978
Bagasse	Cellulomonas sp. Alcaligenes sp.	Batch and continuous	44-50%[a]	Han et al., 1971
	Aspergillus terreus	batch	11.3-29.8% protein	Garg and Neelakantan, 1982
Cellulosic wastes (agriculture, forestry, pulp and paper)	Chaetomium cellulolyticum	Slurry and solid state fermentations	17-40% protein in product	Chahal, 1982 Moo Young, 1983, 1984
Cellulose	Trichoderma viride	Continuous	39-40%[a]	Peitersen, 1977
	Aspergillus terreus	Semicontinuous	78-84% cellulose converted	Miller and Srinivasan, 1983
Mesquite wood	Brevibacterium sp.	Batch	44%[a]	Thayer, et al., 1975

(a) % based on substrate used.

Microbial protein quality includes both nutritional value and functional effectiveness. Important considerations in nutritional value of both food and feed grade protein products include:

(1) "Crude" protein as compared with "true" protein. The former, as determined by multiplying Kjeldahl nitrogen by 6.25 includes nonprotein nitrogen substances such as nucleic acids, which have no nutritional value. The latter is determined by the amino acid content of the protein and is a more realistic measure of protein quality.

(2) Nucleic acid contents. To avoid the possibility of kidney stone formation or gout, nucleic acid contents of food grade products must be reduced so that the total human dietary intake of nucleic acids is less than 2 grams per day. Both enzymatic and ion exchange processes have been developed for removing nucleic acids from microbial proteins (Gierhart and Potter, 1978; Lawford and Lewis, 1982).

(3) Amino acids, vitamin, and mineral contents. For feed grade products, methionine and lysine contents are important since these amino acids are most likely to be limiting. Yeast protein products usuallly provide a full spectrum of B-vitamins. Calcium to phosphorus ratios may have to be adjusted to a 1.2:1 ratio in feeds. It is important that microbial protein products to be used in food not be deficient in iron and zinc. It may be necessary to add a source of these elements in some microbial protein products.

(4) Feeding performance. The most important measure of the value of a microbial protein product for animal feed applications is the performance of the product in actual feeding trials with domestic livestock. Important factors include digestibility, metabolizable energy, weight gain, and efficiency of feed conversion.

Microbial protein products may also have functional properties as food ingredients including water and fat binding, gel formation, dispersing action, whipping and foaming action, coagulability, extrusion and spinning characteristics, and desirable flavours and aromas. Fungal mycelia may be treated to improve textural characteristics for food applications (Simmons, 1982). Some functional food additives derived from microbial cells include bakers yeast protein, bakers yeast glycan, bakers or brewers yeast autolysates and hydrolysates, and yeast extracts. These products have been accepted for food use by regulatory agencies in many countries.

In addition to food and feed grade microbial protein products and functional proteins, there are other food additives such as microbial polysaccharides, acidulants such as lactic, citric, and adipic acids, flavouring substances, sweeteners, and enzymes from biomass-derived carbohydrates that can be made by microbial processes.

Safety is an important consideration with any novel microbial product. These products should be free from plant or microbial toxins, heavy metals, and residues of toxic agricultural or industrial chemicals in addition to acceptable levels of nucleic acids. The Protein Advisory Group of the United Nations (1972, 1974) has published guidelines for assessing the nutritional quality and safety of microbial protein products.

FUTURE DEVELOPMENTS

There are a number of current activities at the research and development stage which could lead to improvements in microbial processes utilizing biomass substrates. These include the following:

(1) Introduction and expression of xylanase activity in yeasts useful for microbial protein production.

(2) Introduction and expression of cellulase activity from cellulolytic microorganisms into *Saccharomyces cerevisiae*.

(3) Development of enlarged-cell yeast mutants for improved recovery by centrifugation.

(4) Development of sensors for in-place measurement of cell mass and glucose in the fermenter during production.

The future for microbial food, feed, and food additive products from biomass will be determined to a large extent by price and availability of available substrates and the quality of the finished product from both nutritional and functional standpoints, assuming that safety criteria set by regulatory agencies are met satisfactorily.

REFERENCES

Bernstein, S., C.H. Tzeng and D. Sisson (1977). The commercial fermentation of cheese whey for the production of protein and/or alcohol. Biotechnol. Bioeng. Symp., 7, 1-9.

Bungay, H.R. (1983). Commercializing biomass conversion. Environ. Sci. Technol., 17, 24A-31A.

Burrell, R.E., W.E. Inniss and C.I. Mayfield (1984). Algal heterotrophy and biomass production. Paper 9.2 Symposium on Biomass Conversion Technology, University of Waterloo, Ontario, July 16-20.

Chahal, D.S. (1982). Bioconversion of lignocellulose into food and feed rich in protein. In "Advances in Agricultural Microbiology", N.S. Suba Rao (ed.), Butterworths, Seven Oaks, United Kingdom, 551-584.

Gadsby, B. and J.A. Simmons(1982). Method of heating and freezing a texturized mycelial fungal mass, U.S. Patent 4,341,806.

Garg, S.K. and S. Neelakantan (1982). Bioconversion of sugarcane bagasse for cellulase enzymes and microbial protein production. J. Food Technol., 17, 271-279.

Gierhart, D.L. and N.N. Potter (1978). Effects of ribonucleic acid removal methods on composition and functional properties of Candida utilis. J. Food Sci., 43, 1705-1713.

Han, Y.W. (1982). Nutritional requirements and growth of a Cellulomonas species on cellulosic substrates. J. Ferment. Technol., 60, 99-104.

Han, Y.W., C.E. Dunlap and C.D. Callihan (1971). Single cell protein from cellulosic wastes. Food Technol., 25, 130-133, 154.

Kristensen, T.P. (1978). Continuous single-cell protein production from Cellulomonas sp. and Candida utilis grown in mixture on barley straw. Eur. J. Appl. Microbiol., 5, 155-163.

Ladish, M.R., K.W. Lin, M. Voloch, and G.T. Tsao. 1983. Process considerations in the enzymatic hydrolysis of biomass. Enzyme Microb. Technol., 5, 82-102.

Lawford, G.R. and P.N. Lewis. 1982. Isolation of microbial protein with reduced nucleic acid content. U.S. Patent 4,330,464.

Lewis, C.W. 1976. Energy requirements for single cell protein production. J. Appl. Chem. Biotechnol., 26, 568-575.

Litchfield, J.H. 1979. Production of single-cell protein for use in food or feed, in "Microbial Technology", 2nd Ed., Vol. 1, H.J. Peppler and D. Perlman, Eds., Academic Press, New York, 93-155.

Litchfield, J.H. 1980. Microbial protein production, Bioscience, 30, 387-396.

Litchfield, J.H.1983a. Single cell proteins, Science, 219, 740-746.

Litchfield, J.H. 1983b. Technical and economic prospects for industrial proteins in the coming decades, in "International Symposium on Single Cell Proteins", J.C. Senez, Ed., Technique et Documentation, Paris, France, 9-33.

Moo-Young, M. 1983. Bioconversion of industrial cellulosic pulp materials to protein enriched product, U.S. Patent 4,379,844.

Moo-Young, M. 1984. Bioconversion of industrial cellulosic pulp materials to protein-enriched product. U.S. Patent 4.447,530.

Miller, T.F. and V.R. Srinivasan. 1983. Production of single cell protein from cellulose by Aspergillus terreus, Biotechnol. Bioeng., 25, 1509-1519.

Peitersen, N. 1977. Continuous cultivation of Trichoderma viride on cellulose. Biotechnol. Bioeng., 19, 337-348.

Peppler, H.J. 1979. Production of yeasts and yeast products, in "Microbial Technology" 2nd Ed., Vol. 1, H.J. Peppler and D. Perlman, Eds., Academic Press, New York, 157-185.

Protein Advisory Group (1972). PAG Guideline No. 12 on Production of Single Cell Protein for Human Consumption" FAO/WHO/UNICEF United Nations, New York.

Protein Advisory Group (1974). "PAG Guideline No. 15 on Nutritional and Safety Aspects of Novel Protein Sources for Animal Feeding" FAO/WHO/UNICEF United Nations, New York.

Solomons, G.L. 1983. Single cell protein. CRC Crit. Rev. Biotechnol., 1, 21-58.

Thayer, D.W., S.P. Yang, A.B. Key, H.H. Yang and J.W. Barker. 1975. Production of cattle feed by growth of bacteria on mesquite wood, Develop. Ind. Microbiol., 16, 465-482.

INDUSTRIAL EXPERIENCE IN COMMERCIALIZATION OF A BIOMASS CONVERSION PROCESS

R.G. McDonald, P.Eng.

Envirocon Canada Inc.
#300 - 475 West Georgia Street
Vancouver, B.C. V6B 4M9

ABSTRACT

An overview of the research and development activities of a commercialization program appropriate for biomass conversion processes is presented. The issues of duration and precedence of the numerous tasks associated with these activities are discussed. In addition, the issues of sources of funding and capabilities required for each of these activities are noted. Details of the case of Envirocon's program to commercialize the Waterloo SCP Bioconversion Process are presented to demonstrate the significance and interdependency of these factors.

INTRODUCTION

Envirocon is an established Canadian biological science and engineering company. Our interests in pollution control have led naturally to the quest for the profitable recovery of commercial products from waste biomass. This, coupled with our strategic planning activities at the end of the 1970's, led to a corporate policy of seeking involvement in the food production field. This policy resulted in aquisition of an international agriculture consulting subsidiary, and establishment of a biotechnology division within the company with the mandate of identifying and commercializing biological processes related to the food industry. To date, investment of serious development effort has been warranted in two of the numerous processes evaluated and tested: organic fertilizer production from municipal sewage wastes and single cell protein production from forestry and agricultural wastes.

This paper provides an overview of the research and development activities, typically applied to commercialization of biomass conversion processes. The case of Envirocon's program to commercialize the Waterloo SCP Bioconversion Process is described and used as a tool to demonstrate the significant issues associated with the commercialization program.

The growing global demand for protein products for animal and human needs has led to an intensive search for unconventional sources. One such increasingly attractive source is cultured microbial biomass, generally referred to as Single Cell Protein (SCP), which can be factory produced by fermentation processes. Envirocon acquired the license for a fermentation process, invented at the University of Waterloo, capable of converting cellulosic wastes into fungal biomass with a protein content approaching that of conventional protein supplements used for animal feed. Envirocon has constructed a pilot plant in Vancouver and conducted process development activities in the plant for a period of 12 months. The total cost of development of this process to date, is approaching $5 million, of which Envirocon has contributed in excess of $2 million; the remainder has been provided by various public sector funding programs.

The Waterloo SCP process is based on the mass cultivation of the fungus, Chaetomium cellulolyticum in a slurried substrate system. The process uses a three stage operation which involves: thermal and/or chemical pretreatment of the cellulosic waste material; aerobic fermentation of the pretreated material with nutrient supplements; and separation of the suspended solids (the product) from the fermentation broth. Cellulosic waste biomass provides the main carbon source for the fermentation. The non-carbon nutrient supplements (nitrogen, phosphorous, potassium, etc.), are derived from chemical fertilizer blends.

THE RESEARCH AND DEVELOPMENT ACTIVITIES

Envirocon's interest in biomass conversion technologies is founded on the approach of marketing the technology itself rather than the products of the technology. With this perspective, the completed technology is the product of interest to Envirocon, and the production of the technology assumes substantial significance. For the purposes of this discussion, the technology production process or commercialization program is segmented into the research phase and the development phase. In practice the boundary between these phases is difficult to identify, rather the technology production process is a continuum, as a result of numerous feedback loops joining the two phases.

As the biomass conversion technology evolves while moving along this continuum, the ideal organization for production of the technology changes as well. In the earliest part of the research phase, the majority of the effort is focussed on basic research and as a result, a research unit specialized in that field is appropriate. During the final stages of commercialization, a process marketing capability is required. Between these two extremes, many varied capabilities are brought to bear. These include process engineering, equipment engineering, economic analysis, patent registration, market analysis, and marketing planning. Associated with these various production inputs, are different costs. Figure 1, displays the relationship between process production cost and stage of development. It is interesting to note that for a typical industrial process the time span is likely to be about ten years and the total cost several million dollars.

The Research Phase

In the case of the Waterloo SCP process research, the majority of the work was carried out in the late 1970's prior to Envirocon's involvement. The fungus Chaetomium cellulolyticum was selected because of its ability to rapidly convert cellulosic substrates directly to protein without the usual hydrolysis-type pretreatments. The research work was carried out in the Department of Chemical Engineering at the University of Waterloo and drew upon the microbiological resources of the University and other institutes.

During the course of the microbiological and process engineering research on the SCP process, the majority of the fermentation tests were conducted in shake-flasks with some additional work done in bench-top stirred tank fermentors. Sufficient new information was generated to warrant application for patents in Canada, the U.S.A. and other countries. Process engineering progressed to the extent that decisions were taken to concentrate on bio-reactor design as a means to improve the process performance.

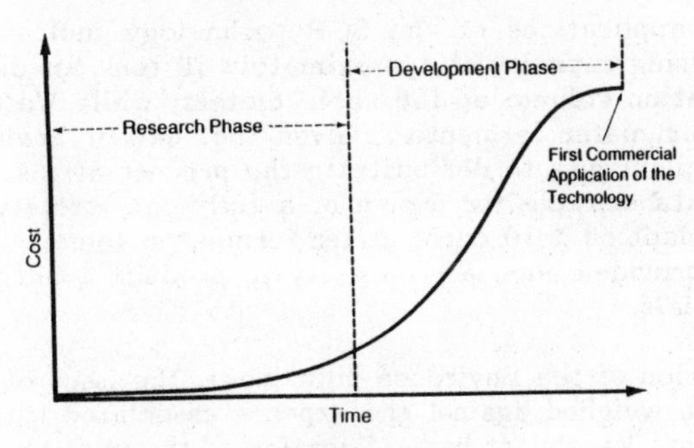

Figure 1
Commercialization of Technology

The Development Phase

It is during the development phase that the first serious evaluation of the market for the product of the process takes place, though this evaluation is preliminary, dealing only with long term trends and the issues of global supply and demand of the product and the raw material of the technology. It is key to the overall success of the program that this study be thorough and objective. The global market for proteins has been thoroughly analyzed and widely published information forecasting the long term growth of the demand for protein is available. Firms such as BP, ICI, Phillips Petroleum and others, have recognized this trend and developed SCP technology as well. The Waterloo technology being based on cellulosic wastes is not as sensitive to changes in oil prices as are the other processes. The Waterloo SCP process easily passes the test of this first market evaluation.

The technical aspects of the development phase include process optimization, pilot scale demonstration, field testing of product, and conceptual design of commercial scale facilities. As a result of their engineering orientation, the Waterloo group was able to undertake many of the tasks associated with these issues.

The transfer of process technology from the research group to the development group, presents numerous challenges to the program team. Communication is the greatest of these since it is not practical to convey much of the relevant information or knowhow in written form. When transferring incomplete technology, a degree of overlap in activities between the two groups is helpful. The research at Waterloo proceeded well into what is normally considered process development. This included assembly of a small scale pilot plant, small scale feeding trials, and conceptual design of full scale plants. It was at this point in the technology development program that Envirocon became involved.

Based on information compiled by in-house economists, Envirocon evaluated the market for the SCP product and SCP production technology in general. Based on information provided by Waterloo, we evaluated the economic and technical viability of their process. Though a substantial amount of development work had been done by Waterloo at that point, Envirocon recognized the need for significant additional investment in time and effort prior to final determination of the commercial viability of the SCP process. To reduce the risk of this investment to an acceptable level, Envirocon sought and obtained funding under the Enterprise Development Program of the Ministry of Industry, Trade and Commerce. The funding provided approximately $1 million toward the design, construction and operation of a pilot plant.

With funding in place, Envirocon assembled a staff of engineers and scientists and together with Waterloo, initiated the transfer of the technology. The most effective vehicle for the transfer of the technology was secondment of Waterloo personnel to Envirocon during the pilot plant design and construction activities.

Analysis of typical applications of the SCP technology indicated that commercial plants will have a processing capacity of approximately 10 tons per day of waste biomass. This dictates a fermentation volume of 100 cubic meters, while Waterloo's pilot plant is equipped with a one cubic meter fermentor. Given that one of Envirocon's major objectives of building a pilot plant was to demonstrate the process at a scale which permitted collection of scale-up data suitable for design of a full scale prototype plant, Envirocon chose to base our pilot plant on a 10 cubic meter fermentor volume. This scale was convenient because it also provided adequate capacity to produce quantities of SCP suitable for large scale feeding trials.

In selecting the location of the Envirocon pilot plant, the issue of convenience of raw material supply had to be weighed against the expense associated with locating scientific staff away from Envirocon's Vancouver base. Location of the pilot plant at a pulp mill site would have provided an on-site demonstration of the technology. However, on balance it was decided that the advantages of having the plant near Vancouver with access to technical and academic resources outweighed the benefits of a pulp mill location. To hedge this decision, Envirocon arranged the pilot plant equipment in structural steel modules which permit easy relocation of the plant should this prove to be appropriate. The plant was installed on the premises of the Western Laboratory of Forintek Canada Inc.

The construction of the pilot plant was completed in the fall of 1982 and the plant was operated for 12 months until September of 1983. Results of these trials confirmed, and in some instances, improved upon the results of the pilot runs carried out at Waterloo. The larger scale equipment resulted in increased growth rate and we have observed rates in excess of 0.3 h^{-1}, as compared to the rate of 0.24 h^{-1} achieved in the Waterloo equipment. Waterloo had forecast crude protein concentrations of about 40 percent, and this level was frequently achieved. Under certain conditions, protein concentrations in excess of 50 percent were obtained.

While pilot operations continued, Envirocon undertook a detailed assessment of the market for SCP technology. This study examined the sources of cellulosic wastes and the demand for protein meals and incentives for SCP production. The results indicated that substrate supply was not a limiting factor, but that only those countries importing protein meals for use in feed formulation, represent attractive prospects for SCP technology. Japan and Western Europe evolved as the best prospects for installation of the first SCP production facilities. Both regions have a large demand for protein meals as a result of their well developed feed industry. Much of this demand is supplied by soymeal imported from the United States and some of these countries offer incentives to increase domestic protein production.

Envirocon also completed the preliminary engineering of a full scale facility based on engineering data obtained from the pilot plant operation. This program included preparation of a capital cost Estimate which is displayed in Figure 2. In 1983 dollars, the capital cost was estimated to be US $3 million for a plant with production capacity of 20 tonnes per day. The results of this capital cost estimate plus the operations data collected in the pilot plant were used as input for a simulation model to arrive at an estimate of US $178 per tonne as the cost of production of SCP. Figure 3 shows the components of this cost which included a cost of US $20 per tonne for handling of the raw material. In some cases, it is likely that this number may be reduced or even become negative if alternative disposal techniques are costly.

The process optimization and development activities were stopped in September of 1983 due to the major technical obstacle of possible product toxicity. Envirocon had been aware of the toxigenic potential of Chaetomium species from near the beginning of our involvement, but it was not perceived to be a serious problem until the results of preliminary feeding trials with chicks indicated significant mortality rate amongst those presented a diet containing SCP. The results were inconclusive, but of sufficient concern to warrant suspension of pilot plant operations until the cause of the problem was identified and resolved. The literature indicates that a number of metabolites of Chaetomium species are possible toxins. Though routine analytical techniques are not available for all of these toxins, we have found concentrations in the order of several hundred parts per billion for some of the toxins in samples of all those batches where feeding trial results were unacceptable.

Recognizing that the toxicity problem would not be resolved at the pilot scale, Envirocon stopped all pilot plant operations and deployed the operations staff to other projects. Three alternate approaches to the toxicity problem have been identified: (1) isolation of a non-toxic strain of C. cellulolyticum, (2) identification of growth conditions (e.g. continuous operation) which will eliminate toxin production, and (3) selection of an alternate fungus to be used as a substitute for C. cellulolyticum. Envirocon is currently initiating programs for each of these approaches in conjunction with the University of Waterloo, Agriculture Canada and the National Research Council.

The timing and outcome of these options are still uncertain, but Envirocon's assessment indicates that technology for conversion of cellulose to protein can provide attractive returns in the longer term. Regardless of the outcome of the toxicity research, Envirocon is determined to apply the physical and human resources assembled for this project to the commercialization of other biomass conversion processes. The combination of our industrial engineering experience and our large scale fermentation pilot plant represent a unique capability in Canada. A capability that will be essential for the successful commercialization of biomass conversion processes currently in the research phase.

FIGURE 2: 1 Full Scale SCP Plant Capital Cost

Preliminary Estimate Expressed in US $1000's at 1983 Prices

Item	Cost
Civil/Structural	$ 170
Mechanical	1,200
Piping	450
Electrical	430
Instrument	85
Subtotal	$2,335
Engineering	$ 220
Contingency and Profit	375
Patent Fees	70
Subtotal	$ 665
TOTAL CAPITAL COST	$3,000

FIGURE 3: SCP Production Cost Summary

Preliminary Estimate expressed in US dollars

Capital Cost: $3,000,000

Production Rate: 6600 tonne SCP per year

Direct Costs	Per Tonne of Product
Chemicals	$ 58.00
Energy	37.00
Labour	25.00
Subtotal	$120.00
Indirect Costs	
Supervision and Administration	10.00
Maintenance	15.00
Laboratory	7.00
Insurance, taxes, etc.	6.00
Subtotal	$ 38.00
Allowance for raw material handling	$ 20.00
PRODUCTION COST	**$178.00**

Section 4

Production and Action of Cellulases

THE CATALYTIC MECHANISM OF CELLULASE

L. Jurasek, M.G. Paice, M. Yaguchi* and S. O'Leary

Pulp and Paper Research Institute of Canada
Pointe Claire, Quebec H9R 3J9

*Div. of Biological Sciences, National Research Council of Canada
Ottawa, Ontario K1A 0R6

This conference has been about the conversion of biomass. When we talk about biomass conversion, most of the time we have cellulase in mind since hydrolysis of cellulose by enzymes plays a prominent role. Thus, there is no wonder that so many papers, including this one, have been devoted to cellulase.

Why should the pulp and paper industry be interested in cellulase? The main purpose of the industry is to preserve cellulose and make use of its unique structural properties. However, there are a number of opportunities for the use of cellulase.

One of the possibilities is to use cellulase to hydrolyze oligosaccharides in spent sulphite liquors and convert them to fermentable sugars, thus increasing yields of ethanol from fermentation of this waste liquor [1]. The pulp and paper industry also produces large volumes of waste fiber in the form of, for example, primary clarifier sludge, which can potentially be converted by cellulase to sugar, a feedstock for various types of fermentation. Thus, cellulase is a key enzyme in the conversion of cellulosic wastes into value-added products. Another possible application which had been experimented with in our Institute is the use of cellulase to digest ball-milled lignocellulosic material and obtain lignin with a very large surface. This lignin has been found to have interesting adsorption properties [2], indicating a possible use as a food additive.

Extensive treatment of cellulosic fibers with cellulase is damaging to the paper strength. However, a limited attack by cellulase restricted to the surface of cellulosic fibers may have the opposite effect. Partial digestion of the fiber surfaces promotes fibrillation, and this results in better bonding of fibers and translates into stronger paper [3]. This process has been labelled as a "biological beating".

Finally, there is interest in cellulase for various fundamental reasons. For example, cellulase is an essential component of the process of wood decay and clarification of its mechanism helps us to understand the problems associated with wood protection. Detailed understanding of the action of cellulase could enable us to develop a useful model for the understanding of the activity of hemicellulases, enzymes which degrade xylan and other hemicelluloses. In some instances, cellulase is also being used as a research tool while investigating the chemical composition of wood and its ultra-structure [4].

The cellulases that have been investigated at Paprican originate mostly from a wood-degrading Basidiomycete Schizophyllum commune. This fungus is a very common saprophyte found on decaying hardwood branches. Its main advantage is that it grows well in a submerged culture and produces large amounts of extracellular enzymes. The fungus has been grown in our laboratory in fermentors and the enzyme production was optimized using a response-surface optimization method [5]. The enzyme yields from the optimized culture are shown in Table 1 [6]. Thus, Schizophyllum commun is an excellent producer of xylanase and beta-glucosidase (cellobiase) and a good producer of CM-cellulase (endo-glucanase), while the avicelase activity, which is largely dependent on cellobiohydrolase, is relatively weak.

Table 1. Enzyme Yields from a Schizophyllum commune Culture Filtrate [6].

Enzyme activity	Yield (I.U./m.)	
	30°C	50°C
CM-cellulase	52.4	-
Filter paper cellulase	0.43	1.32
Avicellase	0.02	-
Cellobiase	17.9	26.5
p-Nitrophenyl- -β-glucosidase	12.5	-
Xylanase	203.0	-

Figure 1 illustrates schematically the ultra-structure of a lignified cell wall and shows how difficult it is for an enzyme to gain access to native cellulose in the lignocellulosic framework. However, if the cellulosic surfaces are made accessible by disintegration of the structure or by the removal of lignin by chemical means, then the process of cellulose hydrolysis can proceed fairly rapidly.

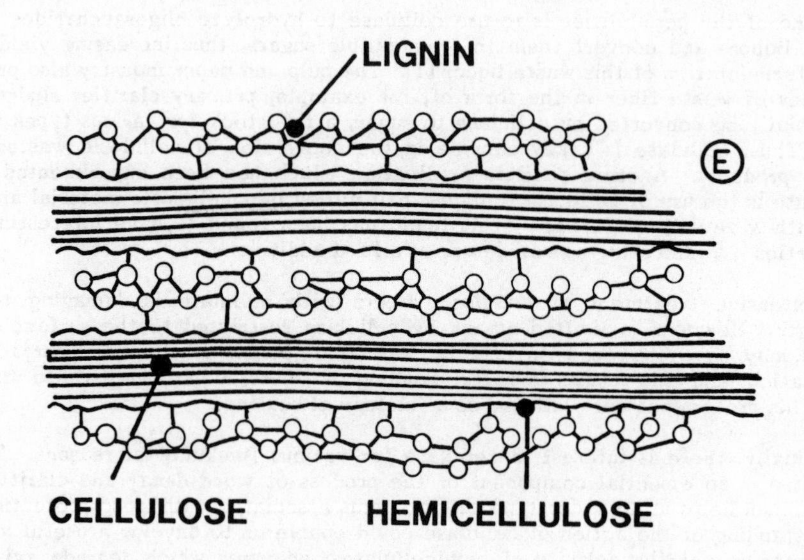

Figure 1. Schematic representation of the ultra-structure of a lignified cell wall [15]. The width of the cellulose strand is about 10 nm and the diameter of an enzyme molecule (E) is approximately 5 nm.

Figure 2 is a schematic representation of the possible functions of the three most important components of the cellulase system [7]. Endoglucanase and cellobiohydrolase work essentially synergistically in a cyclic process, during which cellulose is gradually being stripped off from the surface of the cellulosic microcrystals. The product of this synergistic action is cellobiose, which is converted to glucose by the third component of the system, beta-glucosidase.

Our work has concentrated mostly on the endoglucanase component of the Schizophyllum commune cellulase system, and Figure 3 shows the final purification step of two endoglucanases. In previous steps, the enzymes were purified by a sequence of procedures, including fractional precipitation with ethanol, ultrafiltration and gel filtration.

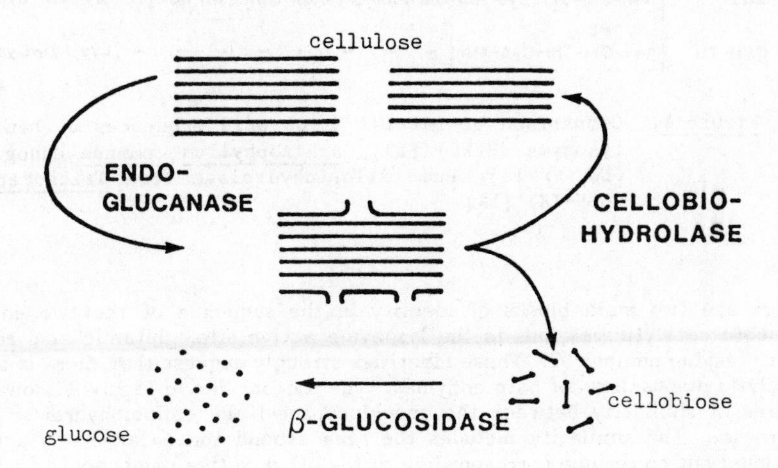

Figure 2. A complex of enzymes converts cellulose to glucose.

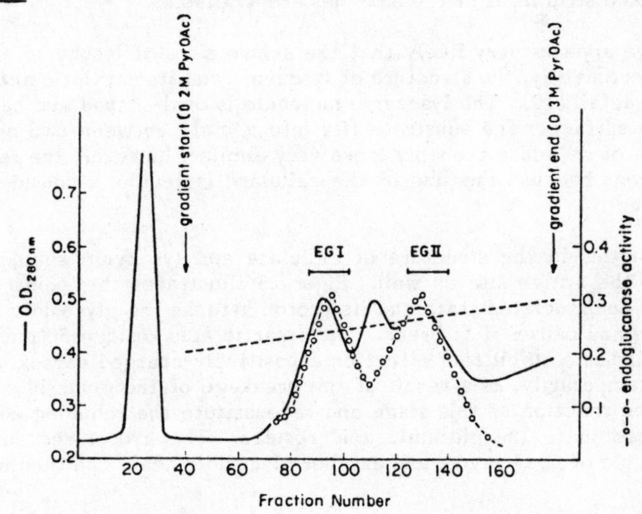

Figure 3. Ion exchange chromatography on DEAE-Bio-gel was used to resolve the two endoglucanases EG I and EG II [8].

The purified endoglucanase-I was then subjected to amino-acid sequence analysis and a partial sequence of the enzyme is represented in Figure 4 [9]. We have attempted to match the sequence of amino acids with a number of enzymes whose structure is known, and the computer analysis revealed a similarity between endoglucanase-I and chicken egg-white lysozyme. This similarity has actually been anticipated because the substrate of lysozyme is quite similar to cellulose in the structure of its backbone. Thus, it is not surprising that lysozyme and cellulase exhibit a certain degree of similarity.

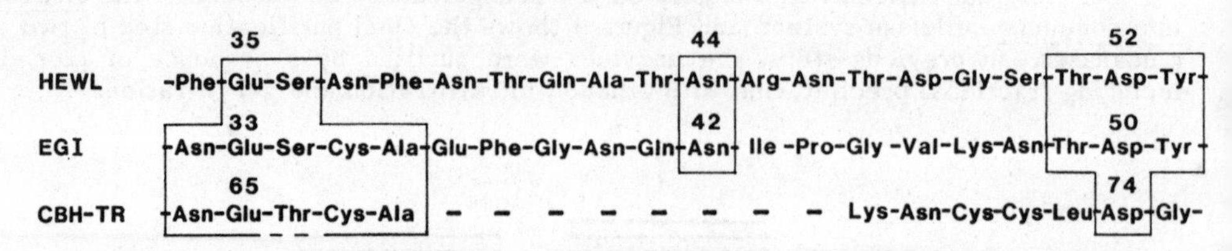

Figure 4. Comparison of partial amino acid sequences of hen egg-white lysozyme (HEWL) [13], Schizophyllum commune endoglucanase-I (EG I) [8], and cellobiohydrolase from Trichoderma reesei (CBH-TR) [14].

There are two main blocks of identity in the sequence of the two enzymes. They include both catalytic residues in the lysozyme active site, glutamic acid residue 35 and aspartate residue number 52. These identities strongly suggest that there is a similarity in the catalytic mechanisms of both enzymes. The bottom line in Figure 4 shows a considerable degree of similarity between the endoglucanase-I and cellobiohydrolase from Trichoderma reesei. The similarity includes the area around the catalytically-active glutamic acid residue; but no residue corresponding to the other active component of active site, the aspartate, has been found in the corresponding position. Thus, an eight-residue deletion was postulated which would bring aspartate residue no. 74 in line with the corresponding residue in the active site of lysozyme. However, this question can only be resolved when full 3-dimensional structures of the enzymes are available.

It therefore appears very likely that the active sites of lysozyme and endoglucanase-I are similar. Fortunately, the structure of lysozyme and its catalytic action has been elucidated in great detail [10]. The lysozyme molecule is oval-shaped and has a deep cleft running across its surface. The substrate fits into a cleft between two active-site residues. The active site of cellulase probably looks very similar; however, the rest of the molecule must be different because the size of the cellulase molecule is considerably greater than that of lysozyme.

The similarities in the structure of cellulase and lysozyme enable us to extend the speculation to the active site as well. Figure 5 illustrates this point. Glutamic acid-33 must be in the undissociated state and its proton attacks the glycosidic linkage of the cellulose molecule and causes it to break. The aspartic acid residue-50 must carry a negative charge, and this has a stabilizing effect on a positively-charged carboxonium intermediate, which arises, temporarily, as a result of the breakage of the glycosidic bond. Water molecules enter the reaction at this stage and reconstitute the reducing end of the cellulosic chain and re-protonate the glutamic acid residue. The hydrolyzed substrate disengages from the molecule of the enzyme and another glycosidic chain can then bind.

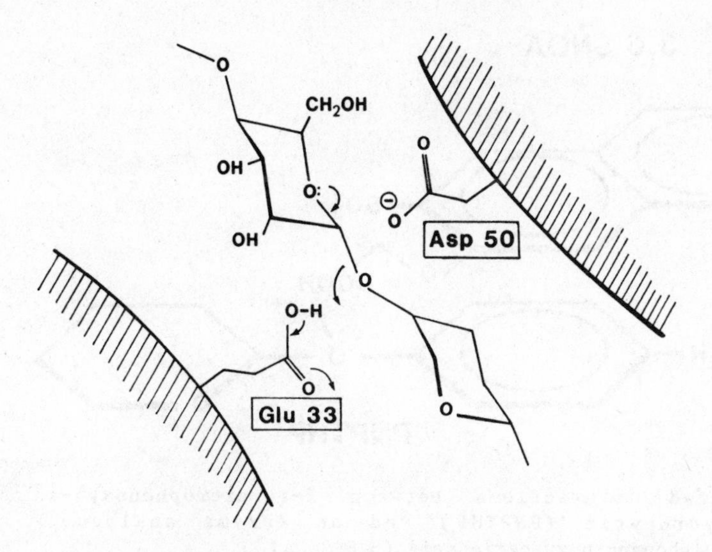

Figure 5. Hypothetical active site of endoglucanase-I and its mode of cleavage of a cellulose molecule [8].

Thus, we have confirmed previous hypotheses [11] about the involvement of an uncharged glutamic acid residue performing general-acid catalysis on the substrate, and of an aspartate residue exerting its effect on the catalytic process by its negative charge. With this hypothesis in mind, we attempted to develop a synthetic enzyme model that would mimic the properties of the enzyme on a much simpler level.

Initial experiments indicated quite clearly that cellulose was very resistant to the process of general-acid catalysis. For this reason, a cellulose model was selected for the experiments [12]. A compound known to be hydrolyzed by general-acid catalysis is 2-(p-nitrophenoxy)-tetrahydropyran (PNPTHP), shown in Figure 6. The acetal linkage joining the two cyclic moieties of this substrate was used as a model of glycosidic linkage. The synthetic enzyme analogue, 3,6 SNOA, was a sulphonylated derivative of naphthoxyacetic acid. It was chosen because its aromatic moiety might interact hydrophobically with the nitrophenyl part of the substrate. Furthermore, the enzyme analogue had a sulphonyl group located in close proximity to its carboxylic-acid gup, to simulate the charge stabilization effected in lysozyme, or cellulase, by the aspartate residue.

Another sulphonyl group was attached to the enzyme model mainly to increase its solubility in water. Figure 6 illustrates the anticipated interaction of the model substrate with the model enzyme, which was expected to enhance the rate of general acid catalysis of the acetal bond.

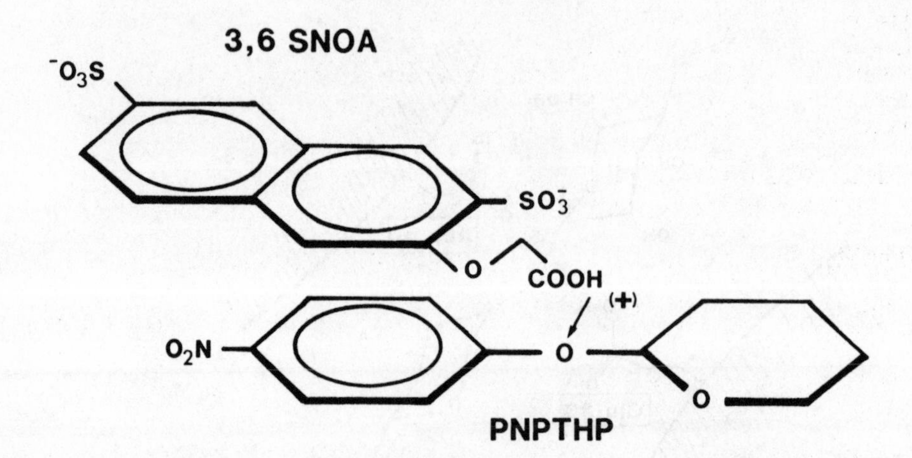

Figure 6. Postulated interactions between 2-(p-nitrophenoxy)-tetrahydropyran (PNPTHP) and an enzyme analogue 3,6-disulphonaphthoxyacetic acid (3,6-SNOA).

This hypothesis is supported by the experimental kinetic data [12]. Figure 7 shows a Bronsted plot, that is, a plot of the logarithm of the second rate constant versus the logarithm of the dissociation constant. The graph was calibrated with three acids which do not exhibit any enzyme-like features. As expected, their catalytic activity is a simple function of pK_a. On the other hand, the synthetic enzyme analogue shows catalytic activity about 30 times higher than should be expected from its pK_a. This rate enhancement is apparently due to a combination of hydrophobic attraction with the substrate, and more importantly, to the presence of a negatively-charged group near the carboxyl group. The contribution of the hydrophobic binding is probably relatively small, as documented by the minor rate enhancement observed in the case of phenoxyacetic acid (Figure 7), which can exhibit hydrophobic interaction, but does not have any negatively-charged groups.

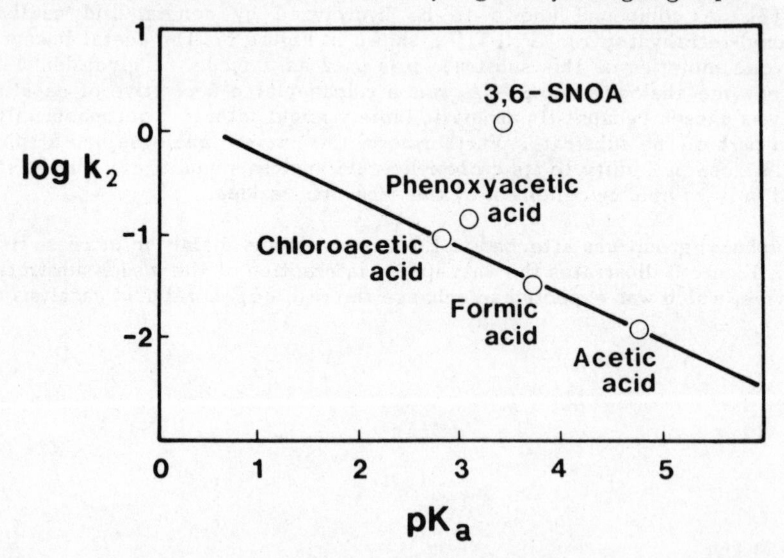

Figure 7. Bronstead plot comparing catalytic behaviour of 3,6-SNOA and phenoxyacetic acid with three "standard" acids [12].

The idea of using synthetic enzyme-like catalysts for cellulose hydrolysis is a very exciting one; however, many hurdles will have to be overcome before it could become of industrial interest. First of all, a catalytic system capable of performing a general-acid catalysis on cellulose will have to be found, and further perfected, to match the enormous enhancement rates observed among natural enzymes.

This is NRCC Publication No. 24085.

REFERENCES

1. M.G. Paice and G. Willick, Proceedings CPPA Tech. Section Annual Meeting, Montreal, A139-143 (1983).

2. G.A. Paden, A.S. Frank, J.M. Wieber, B.A. Pethica, P. Zuman and L. Jurasek, ACS Symposium Series, 214: 241-250 (1983).

3. M.M. Grinberg, V.S. Sivers, V.I. Bilai, Y.V. Lisak and G.V. Koleoneva, USSR Patent 321,563 (1971).

4. M. Sinner, N. Parameswaran and H.H. Dietrich, Adv. Chem. Series, 181: 303-331 (1979).

5. M. Desrochers, L. Jurasek and M.G. Paice, Appl. Environ. Microbiol., 41: 222-228 (1981).

6. M. Desrochers, L. Jurasek and M.G. Paice, Dev. Ind. Microbiol., 22: 675-684 (1981).

7. M. Gritzali and R.D. Brown Jr., Adv. Chem. Series, 181: 237-261 (1979).

8. M.G. Paice, M. Desrochers, D. Rho, L. Jurasek, C. Roy, C.F. Fernand, E. DeMiguel and M. Yaguchi, Bio/Technology, 2: 535-539.

9. M. Yaguchi, C. Roy, C.F. Rollin, M.G. Paice and L. Jurasek, Biochem. Biophys. Res. Commun., 116: 408-411 (1983).

10. C.C.F. Blake, L.N. Johnson, G.A. Mair, A.C.T. North, D.C. Phillips and V.R. Sarma, Proc. Royal Soc. London, B167: 378-388 (1967).

11. M.G. Paice and L. Jurasek, Adv. Chem. Series, 181: 361-374 (1979).

12. S. O'Leary, Can. J. Chem., 621: 1320-1324 (1984).

13. R.E. Canfield, J. Biol. Chem. 238: 2698-2707 (1963).

14. S. Shoemaker, V. Schweickert, M. Ladner, D. Gelfand, S. Kwok, K. Myambo, and M. Innis, Bio/Technology 1: 691-696 (1984).

15. D.H. Page, Wood and Fiber 7: 246-248 (1976).

CLONING OF CELLULASE GENES: GENETIC ENGINEERING AND THE CELLULOSE PROBLEM

J.J. Pasternak and Bernard R. Glick

Department of Biology
University of Waterloo
Waterloo, Ontario
Canada N2L 3G1

The complete utilization of lignocellulosic biomass remains an intractable problem. When food, lumber and paper are processed, vast amounts of waste residues are created which ought to provide, with the appropriate technologies, reliable and economical sources of functional by-products that can be used for the production of food, fuel or chemicals. Although effective, the chemical procedures for fractionating lignocellulosic material into its constituent macromolecular components (lignin, hemicellulose and cellulose) and for subsequent hydrolysis into its essential monomers (phenylpropane, pentoses and hexoses) require corrosion resistant equipment, extensive washing, disposal of chemical wastes, and the input of energy in the form of heat (Fan et al., 1982). Moreover, the overall efficiency of recovery of useful feedstocks is low. With these constraints in mind, attention has centered on enzymatic hydrolysis as a mechanism for the processing of lignocellulose (Brown, 1983). Conceptually, the utilization of enzymatic conversion of lignocellulose seems reasonable. Enzymes with high specificity act rapidly, completely and repeatedly at moderate temperatures (40-50°C). Organisms that produce the degradative enzymes should be able to be grown cheaply, easily and perpetuated indefinitely on the material which is being converted into chemical feedstock. Unfortunately, the perception of potentiality does not always conform to the conditions of the real world. Profound problems due to the nature of the native substrate, the regulatory mechanisms of cellulolytic organisms and the complexity of the cellulase enzyme system per se, have become evident.

Enzymatic degradation of cellulose has been the focus of considerable study. At present, bioconversion of lignocellulose requires extensive chemical and physical pretreatment, and the cellulose that is released is often not a good substrate for enzymatic hydrolysis. As well, cellulolytic organisms have not evolved to carry out large scale industrial processes. An organism is the sum of a complex set of finely controlled reaction systems. If, for example, too much of an end product of biochemical pathway is present the organism can 'sense' this and respond by curtailing its production. Temporary shutting down of a biochemical pathway can be achieved either by inhibiting an enzyme in the pathway (feedback inhibition) or by preventing the production of the enzyme by blocking the synthesis of the mRNA that codes for the specific enzyme (repression, catabolite repression). Other biological constraints (e.g., extent of enzyme secretion, levels of cell density for maximum activity, etc.) lower the overall efficiency of organisms as a means of biodegrading cellulosic biomass. In response to these natural shortcomings, selection and induced mutation programmes have been carried out. To avoid some of the problems of using live material, culture filtrates of cellulolytic organisms have been proposed as a source of cellulolytic activity. Although relatively inexpensive, filtrates yield variable levels of enzymatic activity and are not optimized for cellulose degradation. Finally, although costly and, therefore, impractical at present, the possibility that purified enzymes could be used for efficient breakdown of cellulose has to be considered. The cellulolytic enzymes could be purified from mutant organisms that are not susceptible to feedback inhibition and added to cellulose preparations in proportions that that would give maximum hydrolysis.

The cellulose enzyme systems from various organisms are complex and an understanding of the entire set of reaction mechanisms that are required to completely degrade cellulose is far from complete. A schematic representation of the enzymatic hydrolysis of cellulose, without specifically denoting the status of the substrate, is shown in Fig. 1. The most thoroughly studied pathway for the enzymatic conversion of cellulose to glucose includes endoglucanase (endo-G) which breaks down cellulose chains to oligopolysaccharides, cellobiohydrolase (CBH) which attacks the nonreducing ends of the nicked cellulose chains with the release of cellobiose and cellobiase (β-glucosidase, β-G) which cleaves cellobiose to glucose. An exoglucanase (exo-G) which is distinct from cellobiohydrolase is found in some systems. For the effective hydrolysis of cellulose, all of the components of the cellulase enzyme system must be present in sufficient quantities and in proportions that optimize *in vitro* degradation.

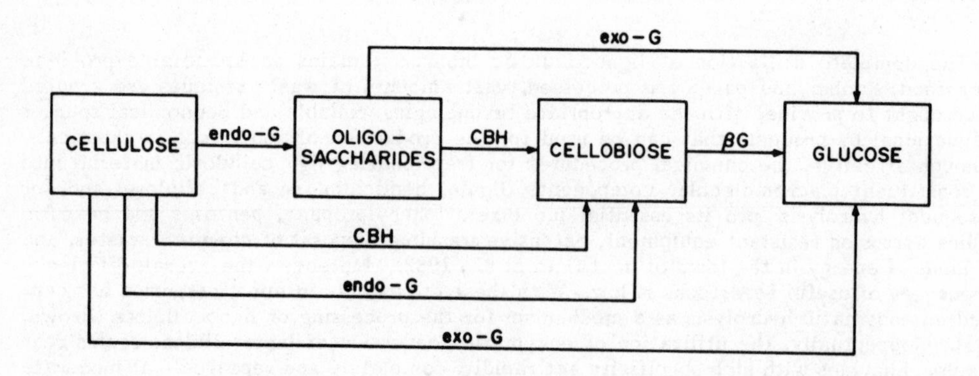

Fig. 1. Schematic representation of enzymatic hydrolysis of cellulose. exo-G, exo-β-1, 4-glucanase; endo-G, endo-β-1, 4-glucanase; CBH, exo-β-1, 4-glucanase (cellobiohydrolase); βG, β-glucosidase (cellobiase).

Thus, what began as a logical and straightforward concept - the enzymatic hydrolysis of lignocellulosic biomass - has, after considerable study, raised a myriad of unforeseen problems. Not unexpectedly, some degree of pessimism concerning the utility of enzymatic systems and/or cellulolytic organisms for effective and economically feasible hydrolysis of lignocellulose has been expressed. Recently, Lutzen et al. (1983) succinctly summarized the current dilemma.

> The use of lignocellulose for the production of ethanol or other chemical feedstocks is one of the most diffucult tasks encountered in the history of biotechnology. Cellulose has a very dense structure and is shielded by hemicelluloses and lignin, so it is very difficult to degrade enzymically. The enzyme systems to be used must contain at least three enzyme components in adequate amounts, which makes production optimization extremely difficult. It is therefore doubtful whether the enzyme production costs can be reduced by one or two orders of magnitude. Consequently it may be necessary to sub-

ject the lignocellulose to a pretreatment before the enzymic hydrolysis, thus increasing the reaction rate to such a degree that the dosage of enzymes can be drastically reduced. Against this background the outlook for enzymic cellulose hydrolysis is rather bleak, and only a dramatic breakthrough can change this picture.

To be sure, a scientific impasse can be overcome by a revolutionary discovery. Alternatively, as noted by W.I.B. Beveridge, scientific advancement often depends on "the adaptation of a piece of new knowledge to another set of circumstances". In this context, the potential of recombinant DNA (rDNA) technology as both a means to produce large quantities of the individual enzymes of the cellulase system and to engineer organisms with specific properties that maximize the biodegradation of lignocellulose must be entertained.

In overview, rDNA technology entails (i) the isolation of a specific DNA fragment (a.k.a insert, passenger or cloned DNA), (ii) the insertion of the passenger DNA into a DNA cloning vehicle (a.k.a plasmid or vector DNA) where the vector has the ability to replicate autonomously in the appropriate host cell, (iii) the introduction of the newly-formed combination of passenger DNA and vector DNA (i.e., hybrid or recombined DNA) into a host cell, (iv) selection and maintenance of those cells that perpetuate the recombined DNA (i.e., transformants) and (v) if possible, the expression (i.e., transcription and translation) of the insert DNA in the host cell (Fig. 2).

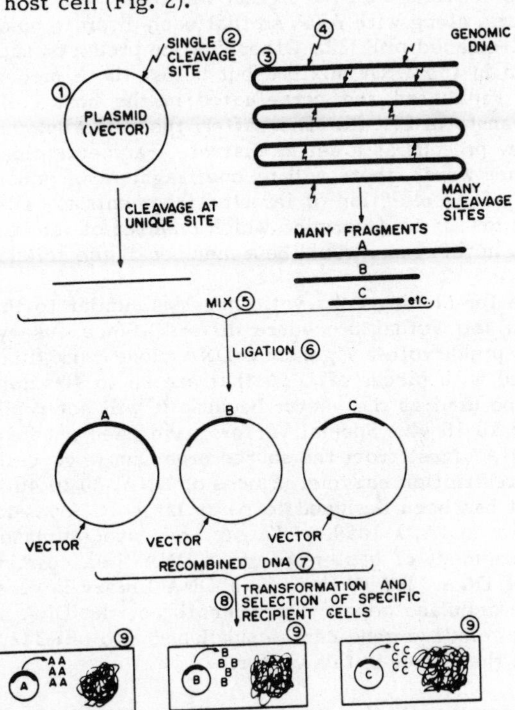

Fig. 2. Generalized scheme for the molecular cloning of genomic DNA into plasmid DNA. Plasmid DNA(1) with a unique restriction site(2) is cleaved. Genomic DNA(3) is also digested the same restriction enzyme(4). The two DNA digests are mixed(5) and treated with ligase(6) to form recombined (hybrid) DNA molecules. A set of recombined DNA molecules(7) are produced. Recipient cells are transformed with the recombined DNA(8). Transformants that express a specific product are selected and cloned(9).

There are two essential features that are required for any successful cloning programme. First, a specific DNA fragment must be inserted into the cloning vehicle. Second, the host cell that receives the vector carrying a specific gene must be readily identifiable. Various strategies have been devised either to identify a specific gene prior to the creation of the recombined DNA or to select a specific transformed cell after it has received the recombined DNA.

The initial requirement of most cloning projects is to generate and maintain discrete pieces of genomic DNA. For prokaryotes, DNA is purified from the source organism (e.g., from an active cellulolytic bacteria) and is digested with a restriction enzyme. The choice of restriction enzyme depends upon the restriction site within the cloning vector into which individual fragments of the partially-digested DNA are to be inserted. Often the restriction enzyme, Sau3A, is used because the pieces it generates can be cloned into a BamH1 site of a plasmid vector such as pBR322. The BamH1 restriction enzyme cleaves DNA at the same sequence as Sau3A except that Sau3A recognizes a sequence of four nucleotides; whereas, BamH1 cleaves a sequence of six nucleotides. The restriction enzyme, Sau3A, will cleave DNA once in every 256 base pairs on the average. A limited digest of genomic DNA will, therefore, yield all possible fragments and a range of sizes. The fragments of genomic DNA (optimum size, 5-10 kilobases) are mixed with pBR322 that has been cut at its unique BamH1 site. The linearized pBR322 is pretreated with alkaline phosphatase to prevent 'self-ligation'. The enzyme, T4 DNA ligase, which can join DNA pieces end-to-end is added to the DNA mixture, along with ATP, so that each discrete piece of genomic DNA is inserted into the BamH1-cleaved pBR322. Other ligation products, such as genomic DNA to genomic DNA can form in the DNA mixture but unless these pieces are inserted into pBR322 they will not be replicated and perpetuated in the host cell. The mixture of ligated DNAs is used to transform E. coli. Thereafter, theoretically, the entire genome of the source organism is now present as a set of discrete fragments cloned into pBR322 in separate host cells. In other words, there will be one fragment of genomic DNA joined to pBR322 per transformant. The collection of inserted fragments is called a clone bank or genomic library. The next task is to determine which member of the bank (library) carries the gene of interest which, in this case, would be a member of the cellulase complex.

Although the rationale for cloning eukaryotic genes is similar to the one that is used for prokaryotic organisms, the actual procedure differs. Since eukaryotic genomes are much larger than those of prokaryotes, a genomic DNA clone bank (library) from a eukaryote should be constructed with pieces of DNA that are 20 to 40 kilobases (kb) long. In this case, pBR322 cannot be used as the vector because it will not replicate well with an insert that is greater than 10-15 kb. Special vectors have been constructed for this purpose. Briefly, a partial DNA digest from the source organism (e.g., cellulolytic fungus) is made with an appropriate restriction enzyme. Pieces of DNA, 20 to 40 kilobases long, are inserted into a vector that has been designed to carry large DNA inserts. Strains of the bacteriophage λ (e.g., λ Charon 4A, λ 1059, λL47, etc.) or a special plasmid cloning vehicle which contains a specific segment of bacteriophage λ DNA (i.e., cosmid) accept and perpetuate large fragments of DNA. In either case, a DNA library is created. This library must then be screened for cellulase genes. Once identified, the DNA segment that contains the information for a cellulase gene can be subcloned into pBR322 or any other suitable vector and maintained thereafter in this vector.

To be effective, a screening procedure must have high resolving power so that only the transformants that carry the specific DNA segment that is being sought is identified from among all the other members of the clone bank. When the insert DNA is from a prokaryote, scoring for the expression of the desired gene is an effective strategy. For example, growth of recipient cells (i.e., members of the clone bank) on cellobiose, when nontransformed cells are incapable of growing in this substrate, was sufficient for isolating a transformant carrying the cellobiase gene (Armentrout and Brown, 1981). To isolate prokaryotic genes which encode endoglucanase activity, the simple, direct and sensitive procedure developed by Teather and Wood (1982) has been used extensively. This method entails plating transformants on a solid selective medium and, after colonies have formed, overlaying them with carboxymethylcellulose (CMC)-agar. After a 16 to 24 hour incubation, the overlay is flooded with a Congo Red solution which is subsequently removed after 30-60 minutes incubation. Colonies expressing endoglucanase activity are identified by the presence of a 'yellowish' halo. This halo represents a region of carboxymethyloligosaccharides, produced by the action of endoglucanase activity, to which Congo Red does not bind. Congo Red preferentially binds to long chain polysaccharides and gives a reddish hue. In this context, Williams (1983) has reported that Grams iodine is effective for scoring halos in a hemicellulose overlay of hemocellulolytic organisms. This approach should facilitate the isolation of the genes which encode hemicellulose-degrading enzymes. Finally, assaying lysates of transformants for CMCase activity directly is tedious but may disclose transformants that synthesize but do not efficiently secrete endoglucanase. Such transformants may not be readily discerned by the 'CMC overlay-Congo Red' method.

A highly specific but complex immunoassay blotting procedure was devised by Whittle et al. (1982) for screening transformants that contained endoglucanase genes from Cellulomonas fimi. This approach provides high resolution and depends upon some expression of the cloned gene. Briefly: A clone bank of C. fimi DNA was generated into the BamH1 site of pBR322. Transformants were selected by insertional inactivation. That is, insertion into the BamH1 site of pBR322 inactivates the tetracycline resistant locus while the ampicillin-resistant gene of pBR322 remains functional so that transformants that carry inserted DNA have an ampicillin-resistant and tetracycline-sensitive phenotype. Bona fide transformants were plated onto solid selective medium. After the formation of colonies, an overlay containing sodium dodecyl sulfate/lysozyme agar, which lysed the uppermost cells of each colony, was applied. Next, CNBr-activated filter paper discs were placed on top of the 'lysis-overlay' agar. The discs bind proteins from each colony. The filter paper discs were then treated with antibody to cellulases from C. fimi, ^{125}I-protein A was applied and the discs were autoradiographed. After developing, spots on the X-ray film denoted those colonies on the original plate that produced C. fimi cellulase proteins. From this plate, live cells from the bottom of positive colonies were retrieved and cultured. This procedure can be used to isolate (i) exoglucanase genes that would give a negative response with the CMC overlay/Congo Red method and (ii) endoglucanase genes whose products are not secreted and therefore not readily scored by the presence of 'Congo Red-negative halos'. In fact, the transformant carrying an endoglucanase gene from C. fimi yields a negative response to the CMC overlay-Congo Red test (Gilkes et al., 1984a).

The strategies for cloning cellulase genes from eukaryotes cannot rely on direct expression in a prokaryotic host cell. Even if yeast, a eukaryote, is used as the host cell, there may be very little expression of foreign eukaryotic genes. However, there is a set of methodologies which do not rely upon gene expression that can be used to clone and identify specific eukaryotic genes. The work of Shoemaker et al. (1983) in isolating the cellobiohydrolase gene (CBH1) of Trichoderma reesei exemplifies such an approach. This strategy relies on the availability of anitbody against cellobiohydrolase. Using oligo-(dT)-cellulose, poly A^+-mRNA was isolated from both cellulase-induced and noninduced Trichoderma cultures. After fractionation of the mRNAs by electrophoresis the mRNA fraction from the cellulose-induced culture that directed the in vitro translation of cellobiohydrolase was identified using antibodies directed against cellobiohydrolase. Complementary DNA (cDNA) was synthesized from this particular mRNA fraction and cloned into pBR322. As well, the comparable mRNA fraction from noninduced cultures which does not contain, to any significant extent, cellulose-induced mRNAs was used to produce cDNA that was also cloned into pBR322. The plasmid DNA from the 'cellulose-induced cDNA-pBR322' clones was used to probe a bacteriophage - T. reesei genomic DNA library by DNA:DNA hybridization. Members of the library that contain DNA sequences that base paired to these cDNA clones were isolated, cultured and then tested by hybridization with plasmid DNA from the non-induced cDNA-pBR322 clones. Any of the initially positive members of the library that did not base pair to the cDNA from noninduced cultures were retained since they should contain DNA sequences that produce mRNA that is specifically transcribed during the induction of cellulases. For experimental convenience, these DNA sequences are routinely subcloned into pBR322 at this stage.

Up to this point, there is no evidence to indicate which, if any, of the positive subclones contain cellulase gene sequences. To answer this critical question, plasmid DNA from positive subclones is isolated, denatured and bound to nitrocellulose filters which are then hybridized with mRNA from cellulose-induced cultures. The mRNA that base pairs to the bound DNA is eluted and used as a template for in vitro translation. The products of each in vitro translation mixture are tested for the presence of newly synthesized material that cross-reacts with antibody against cellobiohydrolase. If this antibody binds to an in vitro translated protein, then the cloned DNA that base paired with the mRNA must contain part or all of the cellobiohydrolase gene.

Using these strategies, for the most part, a number of different research groups have isolated and characterized genes that code for various members of the cellulase complex from diverse cellulolytic organisms. A brief overview on this work is presented here.

(1) Cellobiase. The cellobiase gene from Escherichia adecarboxylata has been cloned into pBR322. After transformation, growth of the recipient E. coli cells on cellobiose was used to screen for the cellobiase gene. The cellobiase is membrane-bound and expressed constitutively in E. coli (Armentrout and Brown, 1981).

(2) Endoglucanase and exoglucanase. Three different cellulase genes have been cloned from Cellulomonas fimi into pBR322 (Whittle et al., 1982; Gilkes et al., 1984a). To enhance secretion of cellulolytic activity, a 'leaky' mutant of E. coli was induced by mutagenesis (Gilkes et al., 1984b). At 40°C, the mutant releases about 40% of the CMCase activity from the periplasmic space. However, the leaky mutant secretes about 1/450 the amount of endoglucanase activity as does C. fimi. Since one of the cellulase clones (pECI) does not yield a positive response with the 'CMC overlay/Congo Red' method, Gilkes et al. (1984a) suggest that the gene may have exoglucanse activity.

(3) Endoglucanase. Two different endoglucanase genes have been cloned from Clostridium thermocellum into E. coli (Cornet et al., 1983a; 1983b). One of the genes (CelA) codes for a 56 kilodalton (KDa) protein and the other (CelB) for a 66 KDa protein. In E. coli, about 62% and 30% of the enzyme activities of the CelA and CelB products are found in the periplasmic space. Neither of the enzymes is secreted into the medium. The level of expression of CelA in E. coli is about 1/300 that in C. thermocellum.

(4) Endoglucanase. Two different endoglucanase genes have been cloned from the alkalophilic bacteria, Bacillus sp. strain N-4 into E. coli (Sashihara et al., 1984). In the recipient bacteria, about 4% of the total endoglucanase activity of one of the genes is released into the medium, 74% is periplasmic and 22% cytoplasmic. With the other gene, 14% of the activity is secreted, 37% periplasmic and 49% cytoplasmic. Endoglucanase activity from the contributing bacterium is stable at 75°C and functions at pH 10.9.

(5) Endoglucanase (xylanase). An endoglucanase gene from the rumen anaerobe Bacteroides succinogenes has been cloned in E. coli (Crosby et al., 1984). DNA hybridization experiments indicate that there may be as many as four additional genes that are related to the cloned endoglucanase gene. The cloned gene is expressed in E. coli although the enzyme has a different pattern of elution from DEAE-Sepharose when it is synthesized by E. coli than by B. succinogenes. The cloned DNA fragment also expressed xylanase activity. However, while the CMCase activity is membrane-bound, the xylanase is secreted into the medium.

(6) Endoglucanase. An endoglucanase gene from the thermophilic actinomycete, Thermomonospora YX has been cloned into E. coli (Collmer and Wilson, 1983). The gene is expressed in E. coli and even when inserted into an expression vector (ptac) the level of expression is only 1/60 the amount produced by the contributing organism. In E. coli, endoglucanase activity is found in the medium (30%), periplasmic space (30%) and cytoplasm (40%).

(7) Cellobiohydrolase. The cellobiohydrolase gene (CBH1) from two strains of *Trichoderma reesei* have been cloned into *E. coli* (Shoemaker et al., 1983; Teeri et al., 1983). Shoemaker et al. (1983) used a strain of *T. reesei* that produces active cellulase in the presence of glucose. The entire CBH1 gene was sequenced and found to code for 496 amino acids. Except for two minor differences, the amino acid sequence data (Fagerstram et al., 1984) agrees with that derived from the CBH1 gene sequence. The genomic sequence of CBH1 contains two introns and the intron-exon junctions are similar to splice sites in yeast. By partial DNA sequencing, Teeri et al. (1983) established that they had also cloned the CBH1 gene from *T. reesei.* Using a newly devised cloning procedure (intron bypass) Steel et al. (1984) have cloned a full-length version of the CBH1 gene from *T. reesei* that is free of introns.

(8) β-glucosidase. The β-glucosidase gene from *Aspergillus niger* has been cloned into a yeast cosmid shuttle vector autonomously in either *E. coli* or yeast. To score for β-glucosidase activity, transformed yeast cells were grown in medium containing the dye 5-bromo-4-chlor-3-indolyl-β-D-glucopyranoside (Xglu) on which positive transformants appear as blue colonies. After subcloning, yeast cells carrying the β-glucosidase gene from *Aspergillus niger* could grow, albeit very slowly, on cellobiose. The products of three other genes from *Aspergillus niger* could not be detected in yeast cells. Thus, these fungal genes are not readily expressed in a heterologous eukaryotic host cell such as yeast.

(9) Xylanase and xylosidase. Xylan-derading genes have been isolated from *Bacillus pumilus* and cloned into *E. coli* (Fukusaki et al., 1984 and references therein). The xylanase gene (xynA) has been sequenced and in *E. coli* the gene product accumulates in the cytoplasm.

(10) Xylanase. The xylanase gene from *Bacillus polymyxa* has been cloned into pBR322 and expressed in *E. coli* (Sandhu and Kennedy, 1984). The immunoassay screening method devised by Whittle et al. (1982) was used to identify a xylanase-producing recombinant clone. The inserted DNA from *B. polymyxa* is about 8.6 kb long.

The long term objectives of recombinant DNA research on cellulase genes are to learn more about the mechanisms of enzyme action, to study at the molecular level how the cellulase enzymes are regulated, to produce at low cost large quantities of the individual components of the cellulase complex and to create organisms with new sets of capabilities that will be able to facilitate industrial utilization of lignocellulosic biomass. Research in this field has just begun. The initial results have been encouraging, but, obviously, no single breakthrough that would resolve the 'cellulose problem' has been discovered.

If imagination fuels experimentation, then the attributes of an idealized genetically engineered lignocellulose-utilizing organism could be delineated. Such an organism would be free of both feedback inhibition and catabolite repression by either glucose or cellobiose, be able to degrade fully hemicellulose and lignin, could readily use crystalline cellulose as a substrate and secrete high levels of the cellulase complex into the medium in amounts for optimal hydrolysis, be capable of both saccharification and fermentation, fuctional at moderate temperature, a nonfastidious grower and tolerant to high alcohol concentrations, among other things.

The rationale for such a dream organism is based, for the most part, on current thinking about how biological utilization of lignocellulose ought to be conducted. One cannot help but get the feeling that researchers sometimes are not unlike the fabled Nasrudin (Shah, 1964) who after losing his housekey looks for it under a street lamp where it is easier to see even though the key was lost in the dark (Fig. 3). The resolution of the 'cellulose problem', in light of Nasrudin's strategy, may well come from an unexpected source. It is not enough to be prepared for this outcome, but, ventures into the 'dark' should be encouraged.

Fig. 3. "People do not always know where to look when they are seeking enlightenment." (I. Shah).

REFERENCES

Armentrout, R.W. and R.D. Brown (1981). Molecular cloning of genes for cellobiose utilization and their expression in Escherichia coli. Appl. Environ. Microbiol., 41, 1355-1362.

Beveridge, W.I.B. (1957). The art of scientific investigation. Random House, New York.

Brown, D.E. (1983). Lignocellulose hydrolysis. Phil. Trans. R. Soc. Lond., B 300, 305-322

Collmer, A. and D.B. Wilson (1983). Cloning and expression of a Thermomonospora YX endocellulase gene in E. coli. Bio/Technology, 1, 594-601.

Cornet, P., D. Tronik, J. Millet and J-P. Aubert (1983a). Cloning and expression in Escherichia coli of Clostridium thermocellum genes coding for amino acid synthesis and cellulose hydrolysis. FEMS Microbiol. Lett., 16, 137-141.

Cornet, P., J. Millet, P. Beguin and J-P. Aubert (1983b). Characterization of two cel (cellulose degradation) genes of Clostridium thermocellum coding for endoglucanases. Biotechnology, 1, 589-594.

Crosby, B., B. Collier, D.Y. Thomas, R.M. Teather and J.D. Erfle (1984). Cloning and expression in Escherichia coli of cellulase genes from Bacteroides succinogenes, Bioenergy Symposium , Paris, France (in press).

Eggling, L. (1983). Lignin - an exceptional biopolymer... and a rich resource? Trends in Biotechnolgy, 1, 123-127.

Fagerstam, L.G., L.G. Peterson and J.A. Engstrom (1984). The primary structure of a 1,4-β-glucan cellobiohydrolase from the fungus Trichoderma reesei QM9414. FEBS Lett., 167, 309-315.

Fan, L.T., Y-H. Lee and M.M. Gharpuray (1982). The nature of lignocellulosics and their pretreatments for enzymatic hydrolysis. Adv. Biochem. Engng., 23, 157-187.

Fukusaki, E., W. Panbangred, A. Shinmyo and H. Okado (1984). The complete nucleotide sequence of the xylanase gene (xynA) of Bacillus pumilus. FEBS Lett., 171, 197-201.

Gilkes, N.R., D.G. Kilburn, R.C Miller Jr. and R.A.J. Warren (1984a). A mutant of Escherichia coli that leaks cellulase activity encoded by cloned cellulase genes from Cellulomonas fimi. Bio/Technology, 2, 259-263.

Gilkes, N.R., D.G. Kilburn, M.L. Langsford, R.C. Miller Jr., W.W. Wakarchuk, R.A.J. Warren, D.J. Whittle and W.K.R. Wong (1984b). Isolation and characterization of Escherichia coli clones expressing cellulase genes from Cellulomonas fimi. J. Gen. Microbiol., 130, 1377-1384.

Lutzen, N.W., M.H. Nielsen, K.M. Oxenboell, M. Schulein and B. Stentebjerg-Olesen (1983). Cellulases and their application in the conversion of lignocellulose to fermentable sugars. Phil. Trans. R. Soc. Lond., B 300, 283-291.

Pentilla, M.E., K.M.H. Nevalainen, A. Raynal and J.K.C Knowles (1984). Cloning of Aspergillus niger genes in yeast. Expression of the gene coding Aspergillusβ-glucosidase. Mol. Gen. Genetics, 194, 494-499.

Sashihara, N., T. Kudo and K. Horikoshi (1984). Molecular cloning and expression of cellulase genes of alkalophilic Bacillus sp. strain N-4 in Escherichia coli. J. Bacteriol., 158, 503-506.

Shah, I. (1964). The Sufis. A Star Book , W.H. Allen Co., Ltd.

Shoemaker, S., V. Schweickart, M. Ladner, D. Gelfand, S. Kwok, K. Myambo and M. Innis (1983). Molecular cloning of exo-cellobiohydrolase 1 derived from Trichoderma reesei strain L27. Bio/Technology, 1, 691-696.

Sandhu, J.S. and J.F. Kennedy (1984). Molecular cloning of Bacillus polymyxa (1→-4)-β-D-xylanase gene in Escherichia coli, Enzyme Microb. Technol., 6, 271-274.

Steel, L.F., T.E. Ward and A. Jacobson (1984). Intron bypass: a rapid procedure for eliminating introns from cloned genomic DNA and its application to a cellulase gene. Nucl. Acid Res., 12, 5879-5895.

Teather, R.M. and P.J. Wood (1982). Use of Congo red-polysaccharide interactions in enumeration and characterization of cellulolytic bacteria from bovine rumen. Appl. Environ. Microbiol., 43, 777-780.

Teeri, T., I. Salovuori and J. Knowles (1983). The molecular cloning of the major cellulase gene from Trichoderma reesei. Bio/Technology, 1, 696-699.

Whittle, D.J., D.G. Kilburn, R.A.J. Warren and R.C. Miller Jr. (1982). Molecular cloning of a Cellulomonas fimi cellulase gene in Escherichia coli. Gene, 17, 139-145.

Williams, A.G. (1983). Staining reactions for the detection of hemicellulose-degrading bacteria. FEMS Microbiol. Lett., 20, 253-258.

CELLULASE PRODUCTION AND HYDROLYSIS OF PRETREATED LIGNOCELLULOSIC SUBSTRATES

J.N. Saddler, M.K.H. Chan, M. Mes-Hartree and C. Breuil
Biotechnology Department
Forintek Canada Corp.
800 Montreal Road
Ottawa, Ontario

In all of the processes which hope to produce ethanol and other liquid fuels from lignocellulosic residues using enzymatic hydrolysis, the most expensive step has been shown to be the production of the cellulase enzymes. It is this requirement for an active cellulase complex acting at optimum conditions on a wide range of cellulosic substrates which has prevented the economic viability of bioconversion processes based on enzymatic hydrolysis. The wood decay fungi have proven to be among the most cellulolytic organisms with relatively few cellulolytic bacteria so far identified. An ongoing screening of cellulolytic fungi and bacteria isolated on field trips and available from culture collections indicated that fungi were generally 50-1000 times more hydrolytic than the most active cellulolytic bacteria. This higher activity was mostly due to the greater amount of extracellular protein secreted by the fungi. Most of the bacteria secreted an incomplete cellullase complex which could only hydrolyse highly modified substrates such as filter paper and carboxymethyl cellulose. When the overall hydrolytic activity of the fungal and bacterial systems were compared it was found that the majority of the complexes were deficient in one or more of the endoglucanase, exoglucanase and β-glucosidase components. Apart from the deficiencies that were apparent in the methods for assaying for individual and collective cellulase activities other factors such as the half-life of the enzymes, enzyme regulation, growth of the organism, etc., all significantly affected the way in which the hydrolytic potential of the various microorganisms could be measured. It was apparent that overall hydrolytic activity as measured by filter paper activity was not representative of the efficiency of hydrolysis of pretreated lignocellulosic substrates. The method of pretreatment and the nature of the substrate was shown to greatly influence how readily the cellulose and hemicellulose components were hydrolysed.

INTRODUCTION

One of the alternatives to going the acid hydrolysis route for the production of sugars from lignocellulosics is the use of enzymatic hydrolysis. However, before efficient enzymatic hydrolysis of most biomass materials can be carried out the substrate must first be pretreated so that the relatively recalcitrant lignocellulosic matrix will be more amenable to conversion. Several chemical, physical and biological methods have been advocated by other workers (Table 1). Most chemical methods are active against the two major deterrents to cellulase hydrolysis, cellulose crystallinity and the lignin barrier. The degree of polymerisation (d.p.) and degree of substitution (d.s.) of the cellulose are also affected by the method of chemical pretreatment and can greatly influence the suitability of some supposedly "model" substrates for assaying cellulase activity. The crystallinity and d.p. of potential substrates such as steam exploded wood (Table 2) varies considerably from commercial substrates such as Avicel or Solka Floc. Lignin has been shown to restrict enzymatic and chemical hydrolysis of the cellulose (Millet et al., 1976).

Most chemical methods use swelling agents which penetrate the crystalline as well as the amorphous region of the cellulose component and lead to new crystalline modifications and possible complete solution of the cellulose after unlimited swelling as occurs in organosolv pulping. Although the agents are efficient at increasing the accessibility of the cellulose substrates there are still problems with product separation, chemical recovery and interference of lignin when whole wood is used as the substrate. Other workers (Baker et al., 1975; Andren et al., 1975) have shown that the presence of lignin in the pretreated wood substrate does not directly interfere with enzymatic hydrolysis as once the cellulose is loosened from the lignin matrix enzymatic hydrolysis can occur even though the lignin is still present. Gassing with SO_2 has been shown to disrupt the lignin-carbohydrate association in situ without the selective removal of either constituent (Millet et al., 1976).

Various types of milling have tended to be the preferred methods of physically pretreating lignocellulosic substrates (Tassinari et al., 1982). By reducing the particle size, disrupting the crystalline structure and breaking down the chemical bonds of the long-chain molecules the susceptibility to enzymatic hydrolysis is substantially increased. The goal of different milling methods is to decrease the crystallinity, increase the surface area for the enzymes to work on and to provide a substrate with high bulk density. Using milling as a pretreatment method has the advantage of being relatively substrate insensitive but suffers from the major disadvantages of being energetically unfavourable and lignin remains as a significant inhibitor to enzyme accessibility.

Cellulases are known to be about 100 times less efficient at hydrolysing cellulose as amylases are at hydrolysing a corresponding amount of starch. Both starch and cellulosic conversion processes have many steps in common with the raw material requiring preparation or pretreatment, hydrolysis, fermentation, recovery and concentration of the alcohol and recovery of by-products. The major difference appears to be that starches can be readily treated to enhance the accessibility of the substrate to the enzymes. During cooking the viscosity increases until a temperature is reached where the starch molecules become solubilised. Unless hydrolysis is initiated the larger molecules aggregate (retrograde) forming inter- and intramolecular associations that are not reactive to commercial saccharifying enzymes. We have observed similar reassociations with cellulosic substrates which have been hydrolysed by hydrofluoric acid (HF) (Saddler and Mes-Hartree, 1983). To circumvent this problem starch is gelatinised during the cooking phase and the D.P. is reduced enough to prevent retrograding on cooling. A liquefaction step is then carried out where the polymers are hydrolysed to at least 10-12 dextrose equivalents using acid or alpha amylase. Acid hydrolysis appears to be the preferred method as the reaction occurs at temperatures which allow complete cooking of all the starch granules. Although alpha amylases are used in some processes, the reaction conditions of around 100°C and pH 6-6.5 greatly reduces the half life of the enzymes. After liquefaction, glucoamylases are added and saccharification occurs at 60-65°C, pH 4-4.5, releasing glucose from the non-reducing end of the polymers.

Although enzymatic saccharification of starch has been shown to be commercially viable, problems still occur in the overall process which reflect the similar problems which still have to be fully resolved before the conversion of cellulose to glucose can be considered a viable process (Table 3). Although the pretreatment of the substrate is interrelated with enzymatic hydrolysis, it is the production of the cellulases and the enzymatic hydrolysis itself which have to be significantly enhanced. Previously, we had shown (Saddler, 1982) that numerous factors can effect the hydrolytic potential of different cellulase complexes. Despite significant advances in increasing the productivity and amount of cellulases that can be produced by different cellulolytic microorganisms using mutation and innovative fermentation techniques (Gallo et al., 1978; Hendy et al., 1982), little progress has been made at increasing the specific activity of the cellulases. By considering the advances that have been made in increasing the effectiveness of cellulase hydrolysis and comparing it with established starch hydrolysis methods we hope to show that the specific activity of cellulases will have to be increased before the enzymatic hydrolysis of lignocellulosic residues will be economically viable.

The method that we are currently using to enhance the conversion of wood to its composite sugars is steam explosion (Saddler et al., 1982; Saddler and Brownell, 1982). This involves the steaming of lignocellulosic substrates at a variety of different temperatures, pressures and retention times with a sharp reduction in pressure to expel the material from the vessel. Other workers (Su et al., 1980) have suggested that rapid decompression does not greatly improve the fiber accessibility to the enzymes although the exploded hardwood fibers are well separated and are probably more readily pumped in slurry than fibers that have not been exploded. The steaming does open up the fiber, renders the hemicellulose soluble in hot water and appears to depolymerise the lignin to some extent (Wayman, 1980). The major disadvantage is that although steam explosion greatly enhances the enzymatic hydrolysis of hardwoods and most agricultural residues tested it has not yet been successfully developed for use with softwoods. To develop an efficient process for converting lignocellulosic material to liquid fuels using enzymatic hydrolysis it is necessary that, the substrate first be pretreated, a highly efficient cellulase system is produced cheaply which is active on the pretreated substrates, all of the cellulose and hemicellulose is hydrolysed to sugars and that this sugar solution can be readily fermented to ethanol and other liquid fuels.

Although pretreatment greatly influences the efficiency of enzymatic hydrolysis the production of the cellulases and the actual hydrolysis itself have been shown to be the most important components of the overall process. Enzyme production costs have been calculated (Wilke et al., 1976) to account for as much as 60% of the total processing costs associated with enzymatic hydrolysis of cellulose to glucose. Although numerous fungi can degrade native cellulose, comparatively few bacteria can degrade soluble cellulose derivatives such as carboxymethyl cellulose while even fewer can extensively degrade insoluble cellulose to sugars <u>in vitro</u> (Mandels, 1975). This is what is required however if a process using enzymatic hydrolysis of lignocellulose to obtain free sugars is to be economically attractive. To achieve complete hydrolysis of the insoluble cellulose substrate, the different enzyme components of the cellulase enzyme complex must be present in the right amounts under the right conditions. This requirement for the synergistic action of different endo- and exo-β-glucanases (cellulases (1,4-(1,3;1,4)-β-D-glucan 4-glucano-hydrolase, EC 3.2.1.4)) andβ-D-glucosidases (β-D-glucosidase glucohydrolase, EC 3.2.1.21) enzymes (Figure 1) for complete cellulose hydrolysis to take place has been reported previously (Wood, 1975); Ryu and Mandels, 1980). It is this requirement for an active cellulase complex acting at optimum conditions combined with the wide range of cellulosic substrates on which it must act which makes efficient enzymatic hydrolysis of cellulosic substrates a difficult problem.

Materials and Methods

<u>Microorganisms</u>. All of the fungi were taken from the Forintek Culture collection (Saddler, 1982). A mycelial inoculum was used to initiate growth in shake flasks containing 150 mL Vogel's medium (Montenecourt and Eveleigh, 1977) along with a designated amount of Solka floc BW 300, glucose or lignocellulosic material. Fungal cultures were harvested when peak cellulase activity was reached by filtration through a Whatman glass fiber filter.

<u>Substrates</u>. Aspen wood chips were saturated with steam at 560 psi for 60 sec., then steam exploded (SE). Some of the pretreated residue was further extracted at room temperature for 2 hours with water (SEW). A portion of this material was subsequently treated for 1 hour with 0.4% NaOH, washed thoroughly with water, mildly acidified with dilute H_2SO_4 and again washed until samples were neutral (SEWA). This procedure and the composition of the pretreated substrates have been described previously (Saddler et al., 1982; Saddler and Brownell, 1982). Solka floc was purchased from Brown and Co., Berlin, N.H., USA. Larchwood xylan was obtained from Sigma.

Assays. Soluble protein was determined directly after TCA precipitation using the modified Lowry method (Lowry et al., 1951) of Tan et al. (1984), using bovine serum albumin (Sigma) as standard. Total reducing sugars were estimated colorimetrically using dinitrosalicylic acid reagent (Miller, 1959). Glucose was determined colorimetrically by the Glucostat enzyme assay (Raabo and Terkildsen, 1960).

Endoglucanase activity (1,4-β-D-glucan 4-glucanohydrolase, EC 3.2.1.4) was determined by incubating 1 mL culture supernatant with 10 mg carboxymethyl cellulose (Sigma; medium viscosity, D.P. 1100, D.S. 0.7) in 1 mL 0.05 M sodium citrate buffer, pH 4.8 at 50°C for 30 min. The reaction was terminated by the addition of 3 mL of dinitrosalicylic acid reagent. The tubes were placed in a boiling water bath for 5 min. then cooled to room temperature and the absorbance read at 575 nm.

Filter paper activity was determined by the method of Mandels et al. (1976). One mL of culture supernate was added to 1 mL of 0.05 M citrate buffer, pH 4.8, containing a 1 cm x 6 cm strip (50 mg) Whatman No. 1 filter paper. After incubation for 1 hr at 50°C the reaction was terminated by the addition of 3 mL dinitrosalicylic acid reagent.

β-glucosidase activity (EC 3.2.1.21) was determined by incubating 1 mL of culture supernate with 10 mg salicin (Sigma) in 1 mL of 0.05 M citrate buffer, pH 4.8, at 50°C for 30 min. The procedure was the same as for the endoglucanase assay.

Xylanase activity was determined by incubating 1 mL of culture supernate with 10 mg of larchwood xylan (Sigma) in 1 mL of 0.05 M citrate buffer (pH 4.8) at 50°C for 30 minutes. The procedure followed was the same as for the endoglucanase assay.

One unit of activity was defined as 1 μmol of glucose equivalents or xylose equivalents released per minute.

Saccharification studies. The cellulosic substrates at a final concentration of 5% cellulose in the total reaction mixture were introduced into 60 mL Wheaton serum vials, along with 5 mL of distilled water. After autoclaving the vials, a 15 mL volume of each enzyme preparation, adjusted to pH 4.8 was added. The amount of reducing sugar and glucose released in each system was measured after 1, 2 or 5 days incubation at 45°C on a New Brunswick orbital shaker, 150 rpm.

Results and Discussion

Currently, a full spectrum of activities are concerned with trying to convert lignocellulosic residues to fermentable sugars. These range from groups hoping to establish pilot plants to those which are trying to clone cellulase genes into fermentative microorganisms. The most cellulolytic organisms so far identified still appear to be the wood degrading fungi (Table 4). As previously mentioned, there are numerous fungi which are capable of hydrolysing native cellulose. This is probably a reflection of the environment from which they are usually isolated, i.e., decomposing biomass, and the fact that their mycelial mode of growth is advantageous when trying to penetrate and excrete enzyme locally against cellulosic substrates. The relatively few bacteria which can hydrolyse native cellulose (Table 5) have a considerably lower hydrolytic potential and have yet to be shown capable of completely hydrolysing lignocellulosic residues at substrate concentrations in excess of 5%. When a comparison between some of the better known cellulolytic fungi and bacteria was carried out (Table 6), expressing the cellulase activities per mg of soluble protein detected in the respective culture filtrates, it can be seen that the various cellulase activities were comparable. Previously, other workers (Ng and Zeikus, 1981) had shown that *Clostridium thermocellum* LQR1 and *Trichoderma reesei* QM9414 had similar endoglucanase, filter paper and xylanase activities when they were compared on the basis of mg of extracellular protein in the culture filtrates. It would appear therefore that fungi are more cellulolytic than bacteria primarily because they produce and secrete more cellulases.

This would also seem to be the basis of most of the recent attempts at increasing the hydrolytic potential of mutated and newly isolated cellulolytic fungi. Most reports to date seem to describe increased cellulolytic potential as either an increase in the strains ability to hydrolyse filter paper, i.e., filter paper activity, IU/mL, or as an increase in the productivity of the strain, i.e., IU/L/h (Table 7). Significant progress has been made in this line of work such that some cellulolytic fungi can be grown on a 5% cellulosic substrate containing an inorganic nitrogen source and produce 2% extracellular protein which is reported to be predominantly cellulases (Tangnu et al., 1981). However, these high concentrations of extracellular protein are probably among the highest yields that have been reported for any microbial based process for producing enzymes and it is unlikely that these values will be increased significantly other than by enhancing production by novel fermentation processes (Hendy et al., 1982, 1984).

Numerous groups are presently trying to increase the productivity of cellulolytic fungi by trying to isolate constitutive cellulase producers. Much of this research is based on the work of Montenecourt and Eveleigh (1979) who isolated the T. reesei Rut C-30 strain which was shown to be a derepressed mutant. Several papers have been published since then claiming higher cellulase yields (Chahal, 1984) and growth on inexpensive media (Warzywoda et al., 1983). More recently, Vandecasteele and Pourquie (1984) have described a T. reesei CL-847 strain which can grow on lactose and produce high levels of cellulase and xylanase activities. Although this mutant still appears to be a derepressed mutant, in that it requires a cellulosic substrate in the media for the cellulases to be produced at an enhanced level, the fact that it produces a full spectrum of activities is significant.

Recently we compared the cellulase activities of T. reesei C30 and T. harzianum E58. A direct comparison was carried out because of the wide range of values that can be obtained depending on the assay procedure followed and the substrate used. Initially, T. reesei C30 and T. harzianum E58 were grown for 8 and 5 days respectively on Solka floc and glucose to determine the constitutive and induced cellulase levels (Table 8). It was apparent that higher extracellular cellulase levels were obtained with T. reesei C30 primarily because of the larger amounts of protein produced by this strain. The lower yields obtained when this strain was grown on glucose confirmed previous reports that this strain is not in fact a constitutive producer but more representative of a de-repressed mutant which can continue to produce cellulases even in the presence of glucose. T. harzianum did not appear to produce as high levels of the cellulase complex as T. reesei , with approximately 50% less protein detected extracellularly. However, the different assays for the cellulase components all indicated that the T. harzianum cellulase complex had a higher specific activity. This strain not only produced higher levels of β-glucosidase, it also released most of the enzymes extracellularly.

As most processes involving enzymatic hydrolysis occur over a longer period of time than is used for the cellulase assays (30-60 mins), culture filtrates of T. harzianum were assayed for their thermal stability at a variety of temperatures (Saddler et al., 1984). At temperatures above 45°C the half life of the various cellulases dropped significantly. Although higher temperatures resulted in higher initial enzyme activities, it was apparent that incubation at elevated temperatures greatly reduced the hydrolytic activity of the cellulase complex. We next incubated culture filtrates of T. harzianum E58 with 5% Solka floc to see what effect different temperatures had on longer term incubation of the enzyme (Table 9). Significantly higher glucose and reducing sugar yeilds were obtained after 48 hrs incubation at 45°C. This indicated that although higher yields could be obtained after an initial 24 hr hydrolysis at 50°C, the half life of the enzyme at elevated temperatures has to be considered for longer term hydrolysis.

We mentioned earlier that the most costly step in the enzymatic hydrolysis of wood is the actual production of the cellulase enzymes. We compared the cellulase activities of T. harzianum E58 and T. reesei C30 after they were grown on a range of substrates (Table 10). T. reesei C30 exhibited traditional regulation, with higher cellulase yields obtained after growth on Solka floc and the steam exploded apenwood (SEW) from which most of the hemicellulose had been removed by water extraction. High xylanase yields were obtained when xylan and the unextracted steam exploded aspenwood (SE) were used as substrates. T. harzianum E58 differed in that it also produced high levels of xylanase when it was grown on the cellulosic substrates.

Culture filtrates of T. harzianum E58 and T. reesei C30 were incubated with a range of substrates at 45°C to see if more realistic lignocellulosic substrates could be as readily hydrolyzed as Solka floc (Table 11). The steam exploded aspenwood which had been water extracted (SEW) was almost as readily hydrolysed as Solka floc and pretreated aspenwood which was further extracted by alkali (SE WA). Although T. reesei C30 was as efficient as T. harzianum E58 at hydrolyzing the substrates to reducing sugars, the latter strain produced significantly higher amounts of free glucose in the hydrolyzate.

It was apparent, therefore, that filter paper activity was not representative of a strain's ability to hydrolyse a realistic lignocellulosic substrate and that a complete cellulase system is required if efficient hydrolysis to glucose is to be obtained. Recently other Trichoderma strains which can produce a full spectrum of activities at high concentrations have been described (Vandecasteele and Pourquie, 1984; Chahal, 1984). However, no increase in specific activity was reported.

Earlier in this manuscript, we discussed the differences and similarities of enzymatically hydrolysing starch and cellulose. One of the major differences was that much more cellulase was required to hydrolyse a specified amount of substrate reflecting the low specific activity of these enzymes. Although the cellulase complex from T. harzianum E58 appeared to have a higher specific activity the increase was not large enough to make a substantial difference in the amount of protein required to completely hydrolyse the substrate. Other increases in the specific activity of cellulases that have been reported (Sheir-Neiss and Montenecourt, 1984) usually show a 2-5 fold increase in the activity of a particular enzyme species, i.e., endoglucanase activity, while the overall hydrolytic potential of the cellulase complex is increased by a significantly lower amount.

In conclusion, it can be stated that progress has been made at increasing the amount of cellulases produced by cellulolytic fungi, however, the actual specific activity has not been greatly increased. It would appear that, if cellulose conversion to glucose is ever to be as efficient as starch hydrolysis to glucose, pretreatment methods have to be devised which enhance the availability of the cellulose to the enzymes while the specific activity of the overall cellulase complex has to be greatly increased.

Table 1. Methods for pretreating lignocellulosic substrates to enhance enzymatic hydrolysis.

Physical	Chemicals	Biological	Combinations
Steaming	Hydrochloric acid	White rot fungi	Steam explosion
Radiation	Sulphuric acid		High temperature milling
Milling	Phosphoric acid		NO_2 + irradiation
Wetting	Sodium hydroxide		Alkali + ball milling
	Ammonia		Sulphur dioxide + steaming
	Sulphur dioxide		

From Dunlap (1980) and Saddler and Brownell (1982)

Table 2. Crystallinity index and degree of polymerization of some cellulosic materials.

Cellulosic Materials	Crystallinity Index	Degree of Polymerization
Avicel	80-84	≃235-245
Carboxymethylcellulose	0	≃300-3,000
Filter paper (Whatman #1)	79.0	≃800-1,500
Solka floc	60-70	≃600-1,100
Raw aspen wood	62.3	≃5,300-10,000
Steam exploded aspen wood	70.7	≃1,500-2,500

Table 3. Common difficulties encountered during the enzymatic hydrolysis of cellulose or starch.

1. Failure to fully pretreat or disperse the substrate
2. Formation of unfermentable sugars and/or inhibitors
3. Incomplete saccharification
4. Low efficiency of enzymatic hydrolysis
5. Incomplete use of liberated sugars
6. Diversion of carbohydrate to products other than the desired product.

Table 4. Extracellular cellulase activities of some cellulolytic fungi.

Fungi	Soluble protein (mg/mL)	Endoglucanase activity (IU/mL)	Filter paper activity (IU/mL)	β-glucosidase activity (IU/mL)	Xylanase activity (IU/mL)	References
Aspergillus terrus	0.8	7.18	0.20	0.39	3.91	Szczodrak et al. (1982)
Chaetomium cellulolyticum	ND	0.30	0.10	ND	ND	Chahal et al. (1977)
Fusarium avenaceum	21.0	8.61	ND	ND	11.29	Zalewska-Sobczak & Urbanek (1981)
Neurospora crassa	ND	8.85	1.03	0.40	ND	Rao et al. (1983)
Penicillium funiculosum	ND	25.0	3.50	11.0	ND	Joglekar et al. (1983)
Schizophyllum commune	ND	0.39	ND	ND	1.60	Paice & Jurasek (1977)
Sclerotium rolfsii	5.4	180.0	1.93	20.5	185.0	Sadana et al (1980)
Trichoderma koningii	ND	0.20	0.15	0.06	ND	Ghai (1980)
Trichoderma reesei C30	19.0	150.0	14.0	0.30	ND	Ryu & Mandels (1980)

ND - not determined.

Table 5. Extracellular cellulase activities of some cellulolytic bacteria.

Bacteria	Soluble protein (mg/mL)	Endoglucanase activity (IU/mL)	Filter paper activity (IU/mL)	β-glucosidase activity (IU/mL)	Xylanase activity (IU/mL)	References
Acetivibrio cellulolyticum	0.10	0.07	0.02	<0.001	ND	Saddler & Khan (1980)
Clostridium thermocellum	0.26	2.50	0.06	<0.001	4.80	Saddler & Chan (unpublished data)
Cellulomonas uda	ND	31.70	0.38	0.07	ND	Nakamura & Kitamura (1982)
Cellvibrio vulgarus	ND	6.50	ND	ND	ND	Oberkotter & Rosenberg (1980)
Bacteroides succinogenes	0.04	0.12	ND	0.003	ND	Groleau & Forsberg (1981)
Streptomyces flavogrisens	1.20	26.60	0.44	0.06	ND	Ishaque & Kluepfel (1980)
Thermoactinomyces sp XY	1.70	83.0	ND	0.17	ND	Hagerdal et al. (1978)
Thermomonospora curvata	0.08	6.50	0.06	ND	ND	Fennington et al. (1982)

ND - not determined

Table 6. Comparison of specific activities between fungi and bacteria.

	Endoglucanase activity (IU/mg)	Filter paper activity (IU/mg)	β-glucosidase activity (IU/mg)	Xylanase activity (IU/mg)
Fungi				
Aspergillus terrus	8.98	0.25	0.49	4.89
Sclerotium rolfsii	33.3	0.35	3.79	34.2
Trichoderma ressei C30	7.89	0.74	0.02	ND
T. reesei QM 9414	7.86	0.34	0.08	1.00
Bacteria				
Acetivibrio cellulolyticus	0.70	0.20	<0.01	ND
Streptomyces flavogrisens	22.2	0.36	0.05	ND
Clostridium thermocellum LQ8	9.62	0.23	<0.01	1.85
C. thermocellum LQR1	4.63	0.05	ND	0.53

ND - not determined

Table 7. Cellulase production by mutant strains of T. reesei

Strain	Soluble Protein (mg/mL)	Productivity FPA (IU/L/h)	Specific Activity (IU/mg)		
			CMC	Filter paper	β-glucosidase
QM 6a*	7	15	12.6	0.7	0.04
QM 9414*	14	30	7.8	0.7	0.04
NG 14*	21	45	6.4	0.7	0.03
C 30#	14	68	8.0	0.7	0.35
MCG 77#	10	105	6.5	0.8	0.06
RL-P37+	8	108	46.8	1.3	0.25
CL-847#	22	125	ND	0.8	0.64

* data from Ryu and Mandels (1980)

\# data from Warzywoda et al. (1983)

\+ data from Sheir-Neiss and Montenecourt (1984)

ND - not determined

Table 8. Cellulase activity of Trichoderma harzianum E58 and T. reesei C30 culture filtrates after 5 and 8 days growth respectively on 2% solka floc and glucose.

Trichoderma-strain	Substrate- (20 mg/mL)	Protein- (mg/mL)	Specific activity (IU/mg) Endoglucanase	β-glucosidase	Filter paper
T. harzianum E58	Solka floc	1.5	33.1	1.8	2.2
	Glucose	0.6	3.6	0.4	0.3
T. reesei C30	Solka floc	3.3	15.4	0.2	1.7
	Glucose	1.0	9.9	0.1	0.6

Table 9. Reducing sugars and glucose released when T. harzianum E58 cellulases[a] were incubated with 5% solka floc at various temperatures.

Incubation temperature°C	Reducing sugars (mg/ml) 24 hrs	48 hrs	Glucose (mg/ml) 24 hrs	48 hrs
30	8.3	20.3	3.6	12.8
37	8.4	20.4	9.0	20.8
45	14.1	27.1	10.3	22.5
50	14.8	21.3	11.7	19.2

[a] Culture filtrate from a 5 day old culture of T. harzianum E58 at a protein concentration of 1.25 mg/ml was used.

Table 10. Cellulase and xylanase activities of Trichoderma harzianum E58 and T. reesei C30 grown for 6 days on various lignocellulosic substrates at 1%.

Trichoderma strain	Substrates* (10 mg/ml)	Endoglucanase activity (IU/ml)	β-glucosidase activity (IU/ml)	Filter Paper activity (IU/ml)	Xylanase activity (IU/ml)
T. harzianum E58	Solka floc	64	2.2	2.7	434
	SE	53	1.8	2.3	350
	SE W	66	2.3	2.8	450
	xylan	5	0.2	0.3	240
T. reesei C30	Solka floc	67	0.2	4.5	30
	SE	55	0.1	3.7	218
	SE W	73	0.2	4.5	140
	xylan	9	0.1	0.4	230

* SE - steam exploded aspenwood exposed to saturated steam at 240°C for 60 sec.
SE W - above fraction which was then extracted with water

Table 11. Hydrolysis of various cellulosic substrates by culture filtrates of T. harzianum E58 and T. reesei C30[a] grown on 1% floc.

Cellulase source	Substrates[b] (50 mg/ml cellulose equivalents)	Reducing sugars released (mg/ml)		Glucose released (mg/ml)	
		24 hrs	120 hrs	24 hrs	120 hrs
T. harzianum E58	Solka floc	15.9	26.5	13.8	26.2
	SE	10.4	19.9	8.3	17.7
	SE W	13.6	23.8	11.4	24.8
	SE WA	14.3	25.4	12.8	25.0
T. reesei C30	Solka floc	13.7	24.4	7.0	12.1
	SE	9.8	19.1	4.7	10.3
	SE W	11.2	21.2	6.4	12.0
	SE WA	13.4	23.8	7.6	14.6

[a] Culture filtrates from a 5 day old culture of T. harzianum E58 (protein conc. of 1.2 mg/ml) and 8 day old culture of T. reesei C30 (protein conc. of 3.1 mg/ml) were inclubated at 40°C with the indicated substrates.

[b] SE; SE W; SE WA - steam exploded aspenwood; steam exploded aspendwood which was water extracted, steam exploded aspenwood which was water and alkali extracted.

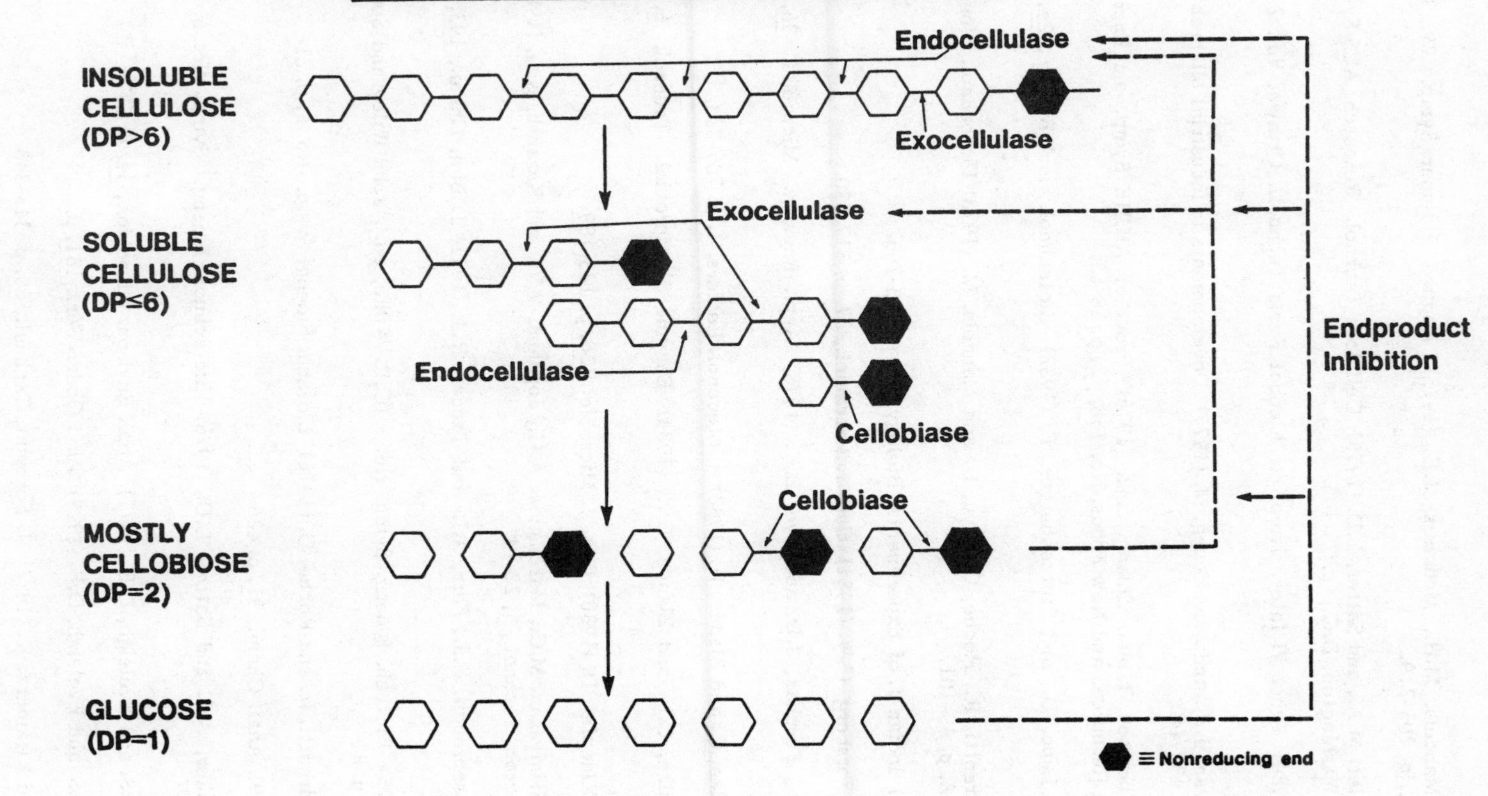

Figure 1. Schematic representation of the synergistic enzyme hydrolysis of cellulose

REFERENCES

Andren, R.K., Mandels, M.H., Medeiros, J.E. (1975) Applied Polymer Symp. 29, T.E. Timell, ed., p. 205-219.

Baker, A.J., Millet, M.A., and Satter, L.D. (1975) Cellulose Technol. Research, ACS Symp. Series 10, Washington, D.C., p. 75-105.

Chahal, D.S. (1984) Proc. VI Inter. Symp. on Alcohol Fuels Technol., Ottawa, Vol. 2, p. 121-127.

Chahal, D.S., Swan, J.E., and Moo-Young, M. (1977) Developments in Industrial Micbrobiology, 18, p. 433-442.

Dunlap, C.E., Thomson, I. and Chiang, L.C. (1976) Proc. of AICHE Symp. on Energy, Renewable Resources and New Foods, No. 158, 72, p. 58-63.

Fennington, G., Lupo, D., and Stutzenberger, F. (1982) Biotechnol. and Bioeng., 24, p. 2487-2497.

Gallo, B.J., Andreotti, R., Roche, C., Ryu, D. and Mandels, M. (1978) Biotechnol. Bioeng. Symp. No. 8, p. 89-101.

Ghai, S.K. (1980) Indian J. of Experimental Biology, 18, p. 703-706.

Groleau, D. and Forsberg, C.W. (1981) Can. J. Microbiol., 27, p. 517-530.

Hagerdal, B.G.R., Ferchak, J.D. and Pye, E.K. (1978) Appl. Environ. Microbiol., 36, p. 606-612.

Hendy, N.A., Wilke, C. and Blanch, H. (1982) Biotechnol. Letters, 4, p. 785-788.

Hendy, N.A., Wilke, C.R. and Blanch, H.N. (1984) Enzyme & Microbiol. Technol., 6, p. 73-77.

Ishaque, M. and Kluepfel, D. (1980) Can. J. Microbiol., 26, p. 183-189.

Joglekar, A.V., Srinivasan, M.C., Marchanda, A.C., Jogdand, V.V., and Karanth, N.G. (1983) Enzyme Microb. Technol., 5, 22-24.

Lowry, O.H., Rosebrough, N.J., Farr, A.L. and Tandal, R.J. (1951) J. Biol. Chem., 193, p. 265- .

Mandels, M. (1975) Biotech. Bioeng. Symp. No. 5 (C.R. Wilke, ed.), John Wiley and Sons, New York, p. 81- .

Mandels, M., Andreotti, R. and Roche, C. (1976) Biotech. Bioeng. Symp. No. 6, p. 21- .

Miller, G.L. (1959) Anal. Chem., 31, p. 426- .

Millet, M.A., Baker, A.J. and Satter, L.D. (1976) Biotechnol. Bioeng. Symp. No. 6, p. 125-154.

Montenecourt, B.S. and Eveleigh, D.E. (1977) Appl. Environ. Microbiol., 34, p. 777- .

Montenecourt, B.S. and Eveleigh, D.E. (1979) Adv. Chem. Ser., 181, p. 289- .

Nakamura, K. and Kitamura, K. (1982) J. Ferment. Technol., 60, p. 343-348.

Ng, T.K. and Zeikus, J.G. (1981) Appl. Environ. Microbiol., 42, p. 231-240.

Oberkotter, L.V. and Rosenberg, F.A. (1980) Proc. 4th Int. Biodeterior. Symp., (T.A. Oxley, D. Allsopp and G. Becker, eds.) Pitman Publ., London, England, p. 148-149.

Raabo, E. and Terkildsen, T.C. (1960) J. Clin. Lab. Invest., 12, p. 402-406.

Rao, M., Deshpande, V., Keskar, S., and Srinivasan, M.C. (1983) Enzyme Microb. Technol., 5, p. 133-136.

Ryu, D.D.Y. and Mandels, M. (1980) Enzyme Microb. Technol., 2, p. 91-102.

Sadana, J.C., Shewale, J.G., and Deshphande, M.V. (1980) Appl. Environ. Microbiol., 39, p. 935-936.

Saddler, J.N. (1982) Enzyme Microb. Technol., 4, p. 414-418.

Saddler, J.N. and Brownell, H.H. (1982) Proc. Int. Symp. on Ethanol from Biomass, Winnipeg, Canada, October, p. 206-230.

Saddler, J.N., Brownell, H.H., Clermont, L.P. and Levitin, N. (1982) Biotechnol. Bioeng., 24, p. 1389-1402.

Saddler, J.N., Hogan, C.M. and Louis-Seize, G. (1984) Manuscript submitted to Eur. J. Appl. Microbiol and Biotech.

Saddler, J.N. and Khan, A.W. (1980) Can J. of Microbiol., 26, p. 760-765.

Saddler, J.N. and Mes-Hartree, M. (1983) in Proc. of 3rd Pan American Symp. on Fuels and Chemicals by Fermentation Antigua, Guatemala, Aug. 1983, p. 104-137.

Sheir-Ness, G., Montenecourt, B.S. (1984) Appl. Microbiol. Biotechnol., 20, p. 46-53.

Su, T.M., Lamad, R.J., Lobos, J., Brennan, M., Smith, J.F., Tabor, D. and Brooks, R. (1980) Final report for U.S. Dept. of Energy for period Dec. 1, 19790Dec. 31, 1980, SERI/TR-8271-1-77.

Szezodrak, J., Trojanowski, J., Ilczuk, Z., and Ginalska, G. (1982) Acta Microbiologica Polonica, 31, p. 257-270.

Tan, L.U.L., Chan, M.K.-H. and Saddler, J.N. (1984) Biotechnol. Letts., 6, No. 3, p. 199-204.

Tangnu, S.K., Blanch, M.W. and Wilke, C.R. (1981) Biotechnol.

MONOCLONAL ANTIBODIES AGAINST CELLOBIOHYDROLASE FROM TRICHODERMA REESEI

F. Riske, I. Labudova *, L. Miller, J.D. Macmillan and D.E. Eveleigh
Dept. of Biochem. and Microbiol., Cook College,
Rutgers, New Brunswick, NJ

* Inst. of Chem., Slovak Academy of Sciences
Bratislava, Czechoslovakia

ABSTRACT

Since polyclonal antibodies to cellobiohydrolase sometimes cross-react with endoglucanases, monoclonal antibodies (McAb) to the exoglucanase were prepared and evaluated with the objective of developing a specific assay for the enzyme. Fusion of myeloma cells with spleen lymphocytes from mice immunized with cellobiohydrolase I (CBH I) from Trichoderma reesei resulted in several hybrid lines that secreted McAb to that enzyme. The McAb from one cell line was studied for its interaction with proteins both in crude culture broth and in purified CBH I from T. reesei using immunoblotting techniques. McAb reacted with the major protein in CBH I (mol. wt. 66 kd, pI 4.2), and also with several additional proteins. These other cross-reacting proteins were not endoglucanases or β-glucosidases and can represent multiple forms or degradation products of CBH I arising through differences in glycosylation or proteolytic modification. Most importantly, the McAb should be useful as a probe for CBH I in a direct assay for this enzyme.

INTRODUCTION

Cellulase is an enzyme complex comprised of three enzymes, cellobiohydrolase (CBH), endoglucanase, and cellobiase. These three enzymes act both synergistically and concertedly to degrade crystalline cellulose substrates. A variety of assay systems have been proposed for defining the activity of both the individual and also the combined activities of these components (Ghose et al., 1981), but no concensus regarding the "best" assay protocols have been reached. A prime example is cellobiohydrolase which may comprise 50-60% of fungal cellulase complexes. Substrates for cellobiohydrolase are also hydrolyzed by endoglucanase and, thus, a specific assay of cellobiohydrolase is not available. Several indirect methods have been proposed for cellobiohydrolase assay including the use of soluble oligosaccharides as substrates (Deshpande et al., 1984; Reese et al., 1967), a linked cellobiose oxido-reductase assay (T. Kelleher, 1983), or the direct monitoring of cellobiose production from cellulose with its attendant difficulties in the presence of endoglucanases (Gum and Brown, 1976). Yet there is still no direct assay for this enzyme. The application of antibodies that are specific for cellobiohydrolase could yield such an assay provided that antibodies can be obtained that show absolute specificity towards this hydrolase.

Immunologic characterization of cellulase components has been recently initiated (Nummi et al., 1980;Schulein et al., 1981). Polyclonal antibodies have been used in defining the endoplasmic reticulum as the site of cellulase synthesis in Trichoderma reesei (Chapman et al., 1983), to characterize two distinct T. reesei cellobiohydrolases (CBH I, CBH II) (Fagerstam and Pettersson, 1979), and also for affinity purification of cellobiohydrolase (Nummi et al., 1983). However, some lack of specificity has been recorded. Polyclonal antibodies developed against cellobiohydrolase I (CBH I), the main cellobiohydrolase produced by T. reesei, cross-reacted with an endoglucanse (endoglucanase I) (Shoemaker et al., 1983), while antibodies raised toward one cellobiohydrolase (Trichoderma viride CBH C) cross-reacted with an endoglucanase (EG IV, Gritzali and Brown, 1979).

Furthermore we found that a polyclonal preparation showed some cross-reactivity to other proteins in T. reesei culture broths in a double diffusion assay (Ouchterlony, 1968). Since we needed antibodies specific to CBH I, it was appropriate to consider making monoclonal antibodies (McAb) to CBH I as they can bind with high affinity to a single site or epitope on the antigen used to prepare them. A McAb capable of binding with an epitope that is only characteristic to CBH I molecules and no others in the cellulase complex or the crude fungal broth, could be used as a reagent to develop a sensitive, highly specific immunoassay system for the enzyme. Such a McAb could also be incorporated into an immunomatrix for affinity purification of CBH I.

Thus, murine hybridomas were prepared and the specificity of the monoclonal antibodies they produced were compared to that of a rabbit polyclonal preparation using immunoblotting techniques.

MATERIALS AND METHODS

Purification of Cellobiohydrolase

Trichoderma reesei Rut C-30 (Cuskey et al., 1983) was grown on a defined salts medium with cellulose as the sole carbon source (Mandels and Weber, 1969). The mycelium was removed by filtration on glass paper, and the culture broth, further clarified by membrane filtration, was then lyophilized to yield crude cellulase. The major cellobiohydrolase (CBH I) was purified by the method of Fagerstam et al., (1977) from the crude enzyme yielding a preparation showing a single band by disc gel electrophoresis. Purification was monitored electrophoretically using denaturing (SDS) PAGE (Anderson et al., 1983), non denaturing PAGE (Brewer et al., 1974) and isoelectric focusing (IEF) (Farkas et al., 1982). Carboxymethylcellulose-agar (CMC) replicas of the non-denaturing PAGE gels were used to determine the presence of contaminating endoglucanase activity using the Congo red procedure (Bartley et al., 1984; Zitomer and Eveleigh, 1982).

Preparation of Hybridoma Cell Lines

Adult female BALB/c mice (Charles River Breeders, Wilmington, MA) were preimmunized with two 250 µl injections, administered intraperitoneally (i.p.) and spaced two weeks apart. Each injection contained 50µg of purified CBH I in complete Freund adjuvant (1:1). The final immunization, performed nine weeks later as described by Stahli et al. (1983), consisted of three daily injections, each containing 100µg CBH I in 200µl of phosphate buffered saline (PBS), administered intravenously on day one and i.p. on days two and three.

On day four, spleen lymphocytes from the immunized mouse were harvested and fused with log phase mouse myeloma cells (line P3-X63-Ag8.8653 obtained from the Institute for Medical Research Camden, NJ) using polyethylene glycol 1000 as the fusigen and the fusion protocol of Siraganian et al. (1983). Cultures of hybridoma cells were maintained in Dulbecco modified Eagle medium (DMEM) supplemented with fetal calf serum (20%), gentamycin (50µg/ml), HEPES (10 mM), and glutamine (2 mM). Cells were cultivated at 37°C in a humidified atmosphere containing 8% CO_2.

Following fusion, one ml amounts of the cell mixture were distributed in the wells of tissue culture plates containing a feeder layer of unfused spleen cells. Hybrid cells were selected using a HAT (hypoxanthine, aminopterin, and thymidine) medium (Littlefield, 1964) prepared with supplemented DMEM as a base.

Cultures producing McAb to CBH I were cloned twice in soft agar (Coffino et al., 1972). McAb were produced in culture broth by growth of cells in 75 cm^2 tissue culture flasks or in ascites fluid following injection of about 10^7 hybrid cells into pristine-primed BALB/c mice.

Detection of Monoclonal Antibodies Against Cellobiohydrolase I

Serum collected from mice three to seven days after the second injection of CBH I, was used for the development of an enzyme-linked immunosorbent assay (ELISA) to detect antibodies against CBH I using the procedures outlined by Voller et al., (1978). Briefly, 100µl PBS, pH 7.4, containing 1 µg of CBH I was added to individual wells in a 96 well polystyrene plate (Nunc, Vanguard Int., Neptune, NJ). After incubation overnight at room temperature, the wells were washed with PBS containing 0.05% Tween 20 , and then filled with 2% bovine serum albumin (BSA) to saturate unbound reactive sites. After 40 min., the wells were washed extensively with PBS-Tween 20, and 100 µl of test hybridoma culture fluid was added and allowed to stand for 45 min. Following another thorough wash, 100 µl of rabbit antimouse IgG-peroxidase was added and incubated for 30 min. at room temperature. The absorbance of the reaction mixtures in each well was measured at 410 nm in an ELISA reader (Biotek, Burlington, Vermont). Cultures were considered positive for production of McAb to CBH I if the absorbance was three to four times higher than that exhibited by reaction of three control McAb prepared to antigens other than CBH I.

Immunoblotting

T. reesei culture broth (Rut C-30) and purified CBH I were analyzed electrophoretically using SDS PAGE, PAGE and IEF tube gels. The separated proteins were then transferred electrophoretically to nitrocellulose paper (Towbin et al., 1979). After blotting, the nitrocellulose paper was soaked in a blocking solution (0.9% NaCl, 1% BSA, 1% gelatin, and 0.02% sodium azide in 0.05 M Tris-HCl, pH 7.4) to saturate unbound sites, washed several times in neutral buffer (0.9% NaCl, 0.1% BSA, 0.1% gelatin, and 0.1% Tween 20 in 0.05 M Tris-HCl, pH 7.4), and sealed in a plastic bag with an appropriate dilution of either McAb or rabbit polyclonal antibody (a gift of Martin Schulein, Novo Corp., Copenhagen, Denmark). After overnight incubation and washing, a peroxidase conjugated ligand (Kirkegaard and Perry Labs Inc., Gaithersburg, MD) specific for mouse IgG was diluted appropriately in neutral buffer and incubated with the paper for two hours. After extensive washing of the paper, the immune complexes were visualized by reaction with 4-chloronaphthol.

RESULTS

Monoclonal Antibodies Against Cellobiohydrolase I

The fusion of spleen lymphocytes from an immunized mouse, with myeloma cells resulted in the generation of a number of hybrid lines. Within 12-15 days after fusion, visible hybridoma growth was detected in 180 of the 240 wells inoculated with the fusion mixture. Culture broth from these wells was tested by ELISA and 44 were found to produce McAb to CBH I. Five cell lines were cloned twice to insure monoclonality and clonal stability (Coffino et al., 1972).

Hybridoma cell line 8B3 was selected for further study and McAb produced by cells grown in tissue culture flasks was purified by affinity chromatography on protein A-Sepharose Cl-4B (Pharmacia Fine Chemicals, Piscataway, NJ) as described by Langone et al., (1982) and analyzed by electrophoresis. Non-denaturing PAGE revealed a single protein migrating in a position equivalent to mouse IgG. SDS-PAGE showed two bands, corresponding to a heavy chain (mol. wt. 54 kd) and a light chain (mol. wt. 31 kd).

Immunoglobulin class and subclass for the five cloned hybrid cell lines were determined by ELISA using a mouse antibody subtype identification kit (Boehringer Mannheim Inc., Indianapolis, IN). Cell line 8B3 secreted a McAb subclass IgG_{2a} and was used for detection of antigen in immunoblotting experiments. One of the other five McAb was an IgG_{2a}, another was an IgG_{2b} and the other two were both IgG_1.

Cellulase Characterization Via Antibody Reactions

(i) Non-denaturing-PAGE. Following electrophoresis of the Rut C-30 culture broth, several protein bands were detected by staining with Coomassie blue (Fig. 1, lane 1). The major protein band migrated to a position equivalent to the major band resolved for CBH I (Fig. 1, lane 2). At least two areas on the gel exhibited endoglucanase activity, visualized via carboxymethylcellulose-Congo red staining (results not shown).

Proteins interacting with either polyclonal or monoclonal antibodies were detected with peroxidase conjugated ligands. More than ten proteins in the Rut C-30 broth reacted with the polyclonal rabbit antisera prepared to CBH I (Fig. 1, lane 3). The purified CBH I preparation yielded one major band for CBH I and several minor bands reacting with the polyclonal antibody preparation (Fig. 1, lane 4). These minor proteins were not detectable by Coomassie blue staining (Fig. 1, lane 2). In contrast, the McAb interacted with CBH I and four additional proteins from the Rut C-30 culture broth (Fig. 1, lane 5). The purified CBH I preparation contained two proteins which separated as close but distinct bands that stained with the monoclonal-peroxidase ligand (Fig. 1, lane 6).

(ii) SDS-PAGE. At least ten proteins in the Rut C-30 broth were resolved and detected by Coomassie blue staining (Fig. 2, lane 2). The major protein, CBH I, migrated to a position corresponding to a molecular weight of about 66 kd (Fig. 2, lane 3). Nitrocellulose blots of these proteins were treated with antibody-peroxidase ligands. Polyclonal antibody reacted with CBH I and in a smear representing at least ten other proteins (Fig. 2, lane 4) while McAb reacted with three distinct proteins, CBH I (66 kd), and proteins with molecular weights of 60 kd and 57 kd (Fig. 2, lane 5).

Nitrocellulose blots of the purified CBH I preparation following electrophoresis revealed that the McAb interacted with a single protein (66 kd) corresponding to the major CBH I band (Fig. 2, lane 7), whereas polyclonal antibody interacted with CBH I and several additional proteins (Fig. 2, lane 5). The amounts of these additional proteins, detected in purified CBH I by the polyclonal antibody, were evidently below the sensitivity of the Coomassie blue method (1 µg) (Fig. 2, lane 3).

(iii) Isoelectric Focusing. Following IEF PAGE in tubes (pH 3.5-10) of the Rut C-30 culture broth, seventeen bands were evident by Coomassie blue staining. At least two proteins (pI 4.05-4.55), including CBH I (pI 4.2), reacted with the CBH I McAb. Appropriate areas of the gels were sliced into 2 mm sections and assayed for enzyme activity. Proteins eluted in the pH range of 4.05-4.55 also contained endoglucanase (CMCase) activity and degraded acid-swollen cellulose, but were not separated sufficiently to permit enzymatic activity to be assigned to specific protein bands. No cellobiase activity was found in this pH range.

In comparison, one major band (pI 4.2) and a minor band (pI 4.1) were detected via McAb in the purified CBH I preparation. These two proteins when eluted were shown to hydrolyze acid-swollen cellulose but not carboxymethylcellulose. Thus, although the purified CBH I preparation yielded two components it was evidently free of contaminating endoglucanase activity by these criteria.

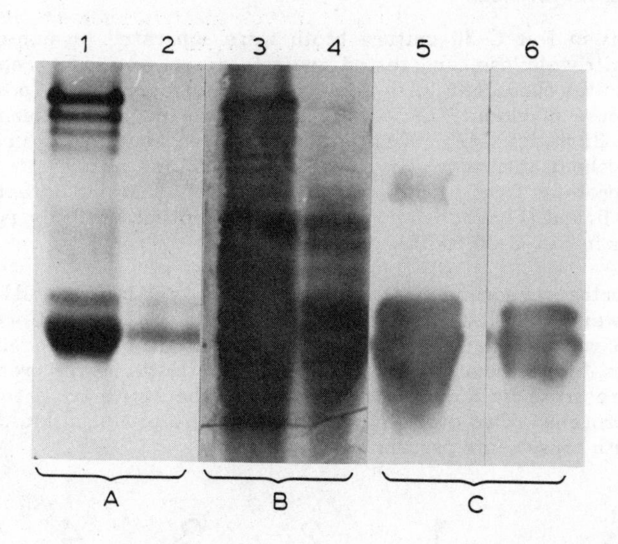

Figure 1

Electrophoretic characterization of cellulase components of T. reesei. Proteins were subjected to electrophoresis under non-denaturing conditions, then stained for protein (Coomassie brilliant blue R-250), or blotted onto nitrocellulose paper and developed with either a polyclonal or monoclonal CBH1 preparation as described under Materials and Methods. (A) Protein stain (B) Nitrocellulose transfer and development with polyclonal antibody (C) Nitrocellulose transfer and development with McAB. Lanes 1,3 & 5 crude culture filtrate: Lanes 2,4 and 6 purified CBH1.

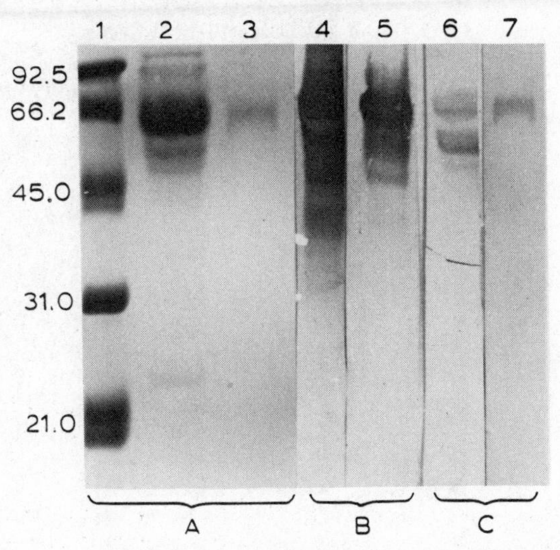

Figure 2

Electrophoretic characterization of cellulase components of T. reesei. Proteins were subjected to SDS-electrophoresis then stained for protein (Coomassie brilliant blue R-250) or blotted onto nitrocellulose paper and developed with either a polyclonal or monoclonal CBH1 preparation as described under Materials and Methods (A) Protein stain (B) Nitrocellulose transfer and polyclonal antibody development (C) Nitrocellulose transfer and monoclonal antibody development. Lane 1, molecular weight standards; lanes 2,4, and 6 crude culture filtrate; lanes 3 and 7 purified cellobiohydrolase 1.

Immunoblotting Controls

Proteins in Rut C-30 culture broth were separated by non-denaturing PAGE, transferred to nitrocellulose, and tested against several antibody preparations (Fig. 3). These included ascites fluid containing McAb against glucose oxidase (prepared in our laboratory), purified mouse myeloma TEPC 183 IgM, purified mouse myeloma protein MOPC 21 IgG_1 (both from Bionetics Corp., Kensington, MD), and McAb from cell line 8B3. The control was PBS without antibody. The McAb from cell line 8B3 reacted with CBH I and a few additional proteins from a total of perhaps twenty proteins detected using the polyclonal-peroxidase ligand (Fig. 3, lane 5). None of the other antibody preparations reacted with components in Rut C-30 broth.

In a further experiment, it was found that McAb from 8B3 did not bind with high molecular weight and low molecular weight standards (Bio-rad Corp., Richmond, California) that had been separated by SDS PAGE and blotted onto nitrocellulose paper. In several experiments, McAb from cell line 8B3 were used to probe for proteins separated from Rut C-30 culture broth by non-denaturing PAGE. The McAb consistently interacted with the same few proteins. Two other hybrid cell lines also produced McAb that exhibited reaction patterns with these same proteins.

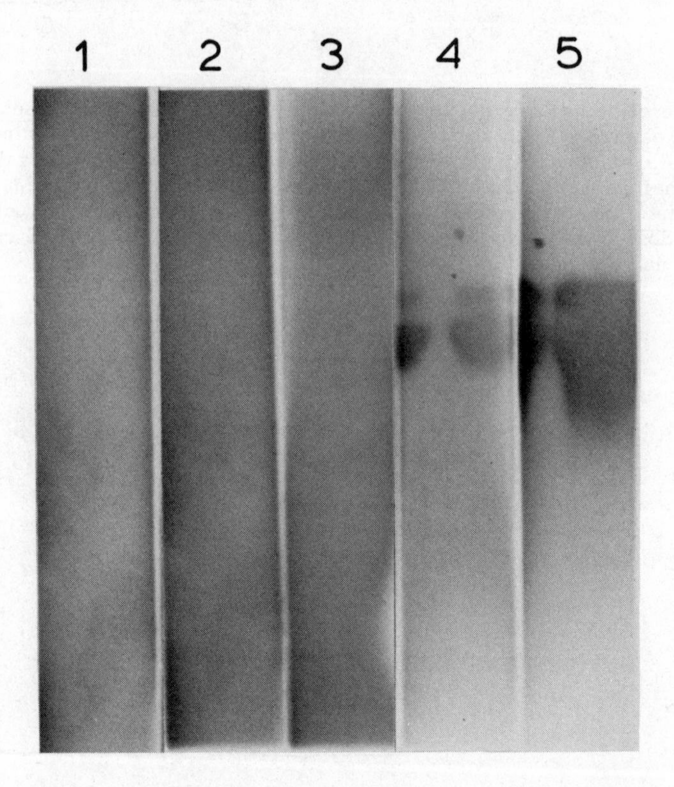

Figure 3

Electrophoretic characterization of cellulase components of T. reesei. Proteins were subjected to electrophoresis under non-denaturing conditions then transferred to nitrocellulose paper, and developed with several different antibody preparations. Lane 1, crude culture filtrate developed with glucose oxidase McAb; lane 2, crude culture filtrate developed with purified mouse myeloma TEPC 183 IgM; lane 3, crude culture filtrate developed with purified mouse myeloma protein MOPC 21 IgG_1; lane 4, purified CBH1 developed with CBH1 McAb; lane 5, crude culture filtrate developed with McAb toward CBH1.

DISCUSSION

The goal of our research was to develop a direct assay for CBH I through the use of antibodies. Initial studies concentrated on defining the specificity of a polyclonal antibody preparation raised in rabbit. Unfortunately, in addition to CBH I, this polyclonal preparation reacted with two partially purified endoglucanases in an Ouchterlony double diffusion assay. Somewhat analogous results have been reported with polyclonal antibodies developed against CBH I cross-reacting in Ouchterlony assays with endoglucanase I (Shoemaker et al.,1983), and antibodies toward CBH C (T. viride) reacting with CBH A,B,C (T. viride), D (T. reesei), and one endoglucanase (EG IV) (Gritzali and Brown, 1979). By combined electrophoresis and immunoblotting, a method of high sensitivity, we found that in addition to CBH I at least ten proteins reacted with the polyclonal antibody preparation (Fig. 1, lane 3; Fig. 2, lane 4). The large amount of cross-reactivity suggested that the polyclonal preparation would not be useful as a specific probe for CBH I.

McAb directed toward a unique epitope on the enzyme would have greater potential for development of a specific assay system. Thus, hybridoma cell lines were generated producing McAb towards CBH I. The McAb from one cell line (8B3) was evaluated as a probe for a direct CBH assay.

The first requirement was to show that the McAb reacted with high sensitivity in probes for CBH I. The electrophoretic position of CBH I in all three systems (SDS, non-denaturing PAGE, and IEF) as visualized with the Coomassie blue stain, corresponded to the major protein recognized by McAb (Fig. 1, lanes 2 and 6; Fig. 2, lanes 3 and 7). Using peroxidase ligands, the McAb was found to react with CBH I when the enzyme was present at levels less than one µg. An apparent mol. wt. of about 66 kd and pI of 4.2 as determined by our work, is similar to values, reported by other investigators (Pettersson et al., 1981; Schulein et al., 1981; Shoemaker et al., 1983). Slight differences in these values have been attributed to varying growth conditions, different culture harvesting times, or the different strains of Trichoderma used to produce the cellulase enzymes (Fagerstam and Pettersson, 1979; Shoemaker and Brown, 1978; Pettersson et al., 1979).

However, the McAb detected a second protein in the purified CBH I preparation (Fig. 1, lane 6). This latter protein was more acidic than the main protein but was only resolved clearly after the bromophenol dye front had migrated off of the gel. IEF separation of the purified CBH I preparation also indicated the presence of a minor protein (pI 4.1) close to the major protein (pI 4.2), which also reacted with the McAb. This minor protein was not CBH II which has a reproted pI of 5.6 - 5.9 (Pettersson et al., 1981) or 5.9 (Shoemaker et al., 1983), and which has been shown to be immunologically distinct from CBH I (Fagerstam and Pettersson, 1979).

Having established McAb interaction with the major CBH I, it was necessary to determine if there was cross-reactivity toward endoglucanases and β-glucosidases. The location of endoglucanases following electrophoretic separation of the Rut C-30 broth as detected using the Congo red assay, did not correspond to any of the proteins which reacted with McAb (results not shown). Similarly, cellobiase activity was not detected in proteins from culture broth in electrophoretic positions corresponding to separated proteins that were reactive with McAb (results not shown). Cellobiase was not present in the purified preparation. Furthermore, the two proteins in the purified CBH I preparation were active on acid-swollen cellulose but not CMC. Therefore, the proteins recognized by McAb in both purified CBH I and the culture broth were not endoglucanases or β-glucosidases.

There is still the point that the McAb cross-reacted with several proteins (from Rut C-30 culture broth) which had pI values in the 4.05 - 4.55 range, MW between 57 - 67 kd, and were resolved into two distinct multiprotein groups according to charge on non-denaturing PAGE. Perhaps analogously, a polyclonal antiserum raised against a purified Avicelase (CBH activity), showed cross-reactivity with at least five proteins, some showing Avicelase activity in crossed immunoelectrophoresis (Fagerstam and Pettersson, 1979).

Similarly, polyclonal antibodies raised toward T. viride cellobiohydrolase C cross-reacted with cellobiohydrolase A and B, and also with T. reesei cellobiohydrolase D (Gum and Brown, 1977). In the study of Schulein et al. (1981), again a polyclonal CBH antibody was found to react with isozymes iso C and iso D. How can these cross-reactions be explained? One explanation for the McAb and the polyclonal interactions with these other proteins could simply be of a nonspecific reaction taking place. This is not likely with the McAb, as other antibody species, known to react with antigens totally unrelated to fungal cellulases, failed to bind proteins separated from Rut C-30 broth (Fig. 3, lanes 1,2,3). This result, coupled with a consistent pattern of McAb (from 3 separate cell lines) interaction with the same few proteins in the Rut C-30 culture broth, indicated that the McAb were reactive with an epitope common to CBH I and these additional proteins.

It is possible that the McAb is reacting with multiple forms of CBH I. These multiple forms could be generated via differential glycosylation of a single polypeptide chain, an explanation strongly supported in one study of such differences of three cellobiohydrolases (A,B, and C) of T. viride (Gum and Brown, 1977). The multiple forms could also arise through proteolytic post-translational changes (Kelleher, 1981; Nakayama et al., 1976). Such glycoproteins would be of reduced molecular size compared to the native enzyme (66 kd) and indeed we have found monoclonal interactive proteins with reduced molecular weights of 57 kd and 60 kd. In our study, the two bands in purified cellobiohydrolase detected by McAb (pI 4.2, pI 4.1) both hydrolyzed acid-swollen cellulose but not carboxymethyl cellulose. Thus, the activitiy of the variant forms was maintained. However, modified proteins need not possess enzymatic activity, but could retain the CBH I epitope recognized by the McAb. The cross-reacting proteins could also be generated artefactually during purification (Nummi et al., 1983).

In summary, the McAb is a probe for CBH I with greater specificity than the polyclonal antibody tested and should be most useful in the development of a direct assay for this enzyme, besides other applications in enzyme purification and cytochemical identification.

ACKNOWLEGEMENTS

New Jersey Agricultural Experiment Station Publication No. F-01111-1-85 supported by the U.S. Department of Energy (DE-AS05-83ER1313140), the U.S. National Academy of Sciences - Czechoslovak Academy of Sciences Exchange Program and New Jersey State Funds.

LITERATURE CITED

Anderson, B.L., R.W. Berry and A. Telser (1983). SDS-polyacrylamide electrophoresis system that separates peptides and proteins in the molecular weight range of 2500 to 9000. Analyt. Biochem., 132, 365-375.

Bartley, T.D., K. Murphy-Holland and D.E. Eveleigh (1984). A method for the detection and differentiation of cellulase components in polyacrylamide gels. Analyt. Biochem., 140, 157-161.

Brewer, J.M., A.J. Pesce and R.B. Ashworth (1974). Experimental techniques in Biochemistry. Prentice-Hall, Englewood Cliffs, NJ.

Chapman, C.M., J.R. Loewenberg, M.J. Schaller and J.E. Piechura (1983). Ultrastructural localization of cellulase in Trichoderma reesei using immunocytochemistry and enzyme cytochemistry. J. Histochem. and Cytochem. 31, 1363-1366.

Coffino, P., R. Baumal, R. Laskov and M.D. Scharff (1972). Cloning of mouse myeloma cells and detection of rare variants. J. Cell Physiol., 79, 429-440.

Cuskey, S.M., B.S. Montenecourt and D.E. Eveleigh (1983). Screening for cellulolytic mutants. In "Liquid Fuel Developments". D.L Wise (Ed.), CRC Press, Boca Raton, FL, 31-48.

Deshpande, M.V., K.-E. Eriksson and L.G. Pettersson (1984). An assay for selective determination of exo-1,4-β-glucanase in a mixture of cellulolytic enzymes. Analytical Biochem., 138, 481-487.

Fagerstam, L.G., U. Hokansson, L.G. Pettersson and L. Andersson (1977). Purification of three different cellulolytic enzymes from Trichoderma viride QM9414 on a large scale. In "Bioconversion of Chemical Substances into Energy, Chemicals and Microbial Protein". T.K. Ghose (Ed.), Thompson Press, Faridabad, Haryana, India, 165-178.

Fagerstam, L.G. and L.G. Pettersson (1979). The cellulolytic complex of Trichoderma reesei QM9414. An immunological approach. FEBS Lett., 98, 363-367.

Farkas, V., A. Jalanko and N. Kolarova (1982). Characteristization of cellulase complexes from Trichoderma reesei QM9414 and its mutants by means of analytical isoelectrofocusing in polyacrylamide gels. Biochim. Biophys. Acta, 706, 105-110.

Ghose, T., B.S. Montenecourt and D.E. Eveleigh (1981). Measure of cellulase activity (substrates, assays, activities and recommendations). Biotechnology Commission, IUPAC, 113 pages.

Gritzali, M. and R.D. Brown Jr. (1979). The cellulase system of Trichoderma. Relationships between purified extracellular enzymes from induced or cellulose-grown cells. In "Hydrolysis of Cellulose: Mechanisms of Enzymatic and Acid Catalysis. R.D. Brown Jr. and L. Jurasek (Eds.), Advances in Chemistry Series 181, American Chemical Society, Washington, D.C., 237-259.

Gum, E.K. and R.D. Brown, Jr. (1976). Structural characterization of a glycoprotein cellulase, 1,4-β-D-glucan cellobiohydrolase C from Trichoderma viride. Biochem. Biophys. Acta, 446, 371-386.

Gum, E.K. and R.D. Brown, Jr. (1977). Comparison of four purified extracellular 1,4-β-D-glucan cellobiohydrolase enzymes from Trichoderma viride. Biochim. Biophys. Acta, 492, 225-231.

Kelleher, T.J. (1981). The lignocellulolytic activity of Phanerochaete chrysosporium Burds. Ph.D. Thesis. Rutgers University, NJ

Langone, J.J. (1982). Applications of immobilized protein A in immunochemical techniques. J. Immunol. Methods 55, 277-296.

Littlefield, J.W. (1964). Selection of hybrids from matings of fibroblast in vitro and their presumed recombinants. Science 145, 709-710.

Mandels, M. and J. Weber (1969). The production of cellulases. In "Cellulases and their Applications", G.J. Hajny and E.T. Reese (Eds.), Advances in Chemistry Series 95, Amer. Chem. Soc., 391-414.

Nakayama, M., Y. Tomita, H. Suzuki and K. Nishizawa (1976). Partial proteolysis of some cellulase components from Trichoderma viride and the substrate specificity of the modified products. J. Biochem. (Tokyo). 79, 955-966.

Nummi, M., M.-L. Niku Paavola, T.-M. Enari and V. Raunio (1980). Immunoelectrophoretic detection of cellulases. FEBS Letters 113, 164-166.

Nummi, N., M.-L. Niku-Paavola, A. Lappalainen, T.-M. Enari and V. Raunio (1983). Cellobiohydrolase from Trichoderma reesei, Biochem. J. 215, 677-683.

Ouchterlony, O. (1968). Handbook of Immunodiffusion and Immunoelectrophoresis. Ann Arbor Science Publisher, Ann Arbor, MI.

Pettersson, G., L. Fagerstam, R. Bhikhabhai and K. Leandoer (1981). The cellulase complex of Trichodrma reesei QM9414. In "The Ekman Day", International Symposium on Wood and Pulping Chemistry, Stockholm. Volume 3, 39-42.

Reese, E.T., A.H. Maguire and F.W. Parrish (1967). Glucosidase and exoglucanase. Can. J. Biochem. 46, 25-34.

Schulein, M., H.E. Schiff, P. Schneider and C. Dambmann (1981). Immunoelectrophoretic characterization of cellulolytic enzymes from Trichoderma reesei In "Bioconversion and Biochemical Engineering" Vol. 1., T.K. Ghose (Ed.), Pub. Pramodh Kapur, Raj Bandhu Industrial Co., C-61 Maya Puri, New Delhi, India, 97-105.

Shoemaker, S.P., and R.D. Brown, Jr. (1978). Enzymatic activities of endo-1,4-β-D glucanases purified from Trichoderma viride. Biochim. et Biophys. Acta. 528, 133-146.

Shoemaker, S., K. Watt, G. Tsitovsky and R. Cox (1983). Characterization and properties of cellulases purified from Trichoderma reesei strain L27. Bio/Technology 1, 687-690.

Siraganian, R.P., P.C. Fox and E.H. Berensyein (1983). Methods of enhancing the frequency of antigen-specific hybridomas. Methods Enzymol. 92, 17-26.

Stahli, C., T. Staehelin, and V. Miggiano (1983). Spleen cell analysis and optimal immunization for high-frequency production of specific hybridomas. Methods in Enzymology 92, 26-36.

Towbin, H., T. Staehelin and J. Gordon (1979). Electrophoretic transfer of proteins from polyacrylamide gels to nitrocellulose sheets: procedure and some applications. Proc. Natl. Acad. Sci., USA 76, 4350-4354.

Voller, A., A. Bartlett and D.E. Bidwell (1978). Enzyme immunoassays with special reference to ELISA techniques. J. Clin. Pathol. 31, 507-520.

Zitomer, S.W. and D.E. Eveleigh (1983). Screening for fungal cellulase mutants through use of Congo red-carboxymethyl cellulose interactions. Abstracts 83rd National Meeting Amer. Soc. Microbiol. (Abstract 068), New Orleans, LA.

CELLULASES FROM A WHITE-ROT FUNGUS: INDUCTION, SECRETION, AND GENE ISOLATION AND FOREIGN EXPRESSION

G.E. Willick, F. Moranelli, and V.L. Seligy

Division of Biological Sciences, Molecular Genetics Section,
National Research Council of Canada, Ottawa, Canada, K1A 0R6

INTRODUCTION

We have been carrying out a program of study on the cellulases of the white-rot fungus, Schizophyllum commune. Our purpose is to understand the biosynthesis and secretion of the cellulases, and to clone and sequence the genes coding for these enzymes. This, taken together with the extensive work carried out on the purification, enzymology, and sequencing of these enzymes should provide a rather unique overall picture of the cellulolytic activity of this organism (1,2,3).

The purpose of this effort is two-fold. Firstly, we wish to express these cellulases in a yeast with the view to creating a commercial yeast capable of directly converting cellulose, or soluble oligosaccharides, into ethanol. Secondly, since these enzymes are very efficiently exported from the fungus, we hope to use the information derived from their study in the development of efficient secretion vehicles for suitable commercially important products produced in yeast.

Our progress in this work has been previously reported (4,5,6,7). Here we shall bring these reports up to date and discuss our current knowledge of the cellulases from S. commune, their biosynthesis and export, and the successful cloning of the endoglucanase from S. commune.

CELLULASES FROM S. COMMUNE

The attack on cellulose by fungi has been shown to involve the cooperative action of three types of 1,4-β-D-glucanases (8), including endo-1,4-β-glucanases (EC 3.2.1.4.), exo-1,4-β-D-glucanases ((EC 3.2.1.91) usually releasing cellobiose), and 1,4-β-D-glucosidases (EC 3.2.1.21) (which may also attack oligosaccharides higher than cellobiose (15)). These enzymes release, respectively, smaller oligosaccharide fragments, cellobiose, and glucose. S. commune secretes at least one member from each class, which have been purified (3).

The enzymes, with their abbreviations, molecular weights, and isoelectric points are listed in Table I. It can be see that they are all very acidic, with the isoelectric points very close together. This, together with their inherent heterogeneity (discussed below), has made their purification difficult.

TABLE I Cellulases purified from S. Commune [a]

Enzyme		Mol wt[b]	pI[c]
β-glucosidase	βGase-I	97,000	3.2, 3.1[d]
	βGase-II	92,000	
Cellobiohydrolase	Av-II[e]	64,000	3.0
	Av-I	61,000	3.15
Endoglucanase	CMCase-Ia	44,000	2.95
	CMCase-Ib	41,000	2.95
	CMCase-II	38,000	2.90

a. Abstracted from ref. 3.
b. Molecular weights estimated by SDS-PAGE.
c. Isoelectric points estimated using a pH 2.5-8.0 gradient.
d. Molecular weight correlation not known.
e. Activity measured against Avicel.

BIOSYNTHESIS AND SECRETION OF CELLULASES

The molecular events in the biosynthesis and secretion of eucaryotic extracellular proteins are broadly known, and are summarized below (Fig. 1).

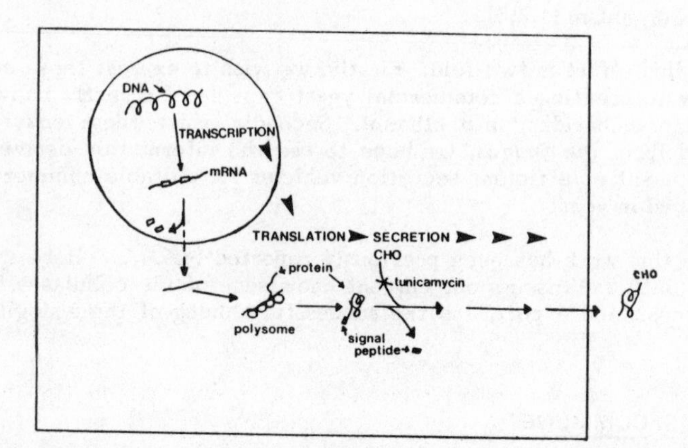

Fig. 1 Biosynthesis and secretion of eucaryotic proteins. The gene is transcribed to give the mRNA, which frequently contains non-protein coding intervening sequences. These are processed out and the mRNA is passed from the nucleus to the cytoplasm. The protein is synthesized on a mRNA-ribosome complex (polysome). Proteins destined to be secreted contain hydrophobic leader sequences (signal peptides) which initiate binding of the protein to a cellular internal membrane (endoplasmic reticulum). The signal peptide is usually cleaved, and the protein passes through secretory organelles, during which time it is glycosylated, and finally passes through the plasma membrane and cell wall into the medium.

Secretion in response to cellulose - S. commune secretes the full complement of cellulases in response to cellulose. When grown on glucose, the organism secretes little or none of the identified cellulases (3). On removing the medium, and replacing with fresh cellulose containing medium, the cellulases are secreted after a lag period (Fig. 2).

We also found that the maximum level of the mRNAs coding for the cellulolytic enzymes, as indicated by their ability to direct cell-free translation of the cellulases in the rabbit reticulocyte system, occurred at about 3 days (Fig. 2). We also noted that the βGase transcript appeared slightly ahead of the CMCase, whereas the Av transcript appeared coincidentally. This agrees with our previous observation that the βGase is not expressed coordinately with the CMCase and Av (3).

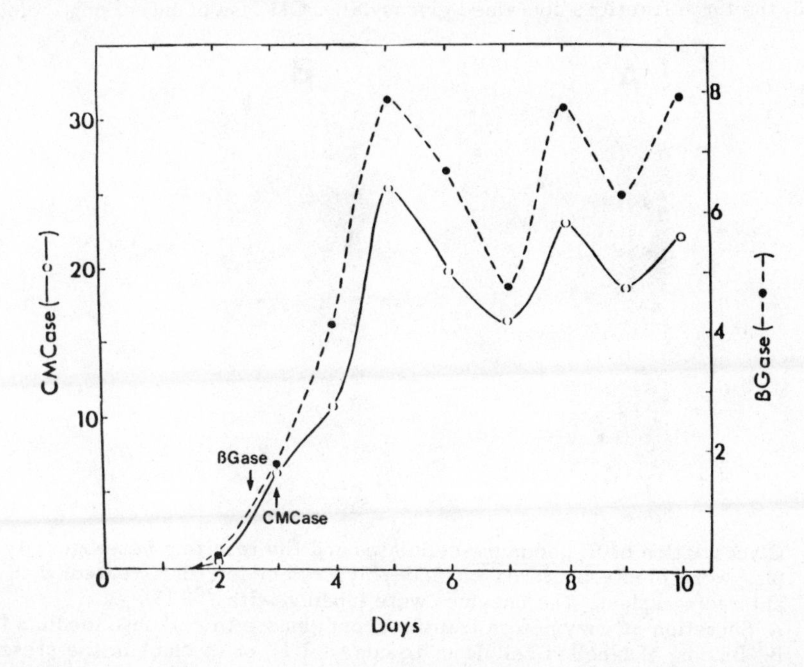

Fig. 2 Induction of cellulase secretion by cellulose. The mycelium was transferred from a glucose to cellulose medium at time zero. The maximum rate of secretion corresponded to the maximum level of cellulase coding mRNA, as demonstrated by their ability to synthesize the cellulases in the rabbit reticulocyte cell free system. The maximums are shown by arrows for the CMCase and βGase.

The results of these experiments indicated that the primary part of the heterogeneity in the enzymes is determined at the gene level. At this point, our data suggests the presence of only one gene coding for the CMCase. This then indicates that the heterogeneity in the transcripts arises either from differential transcription of the same gene, or a differential processing of the resulting mRNA. Examples of each case have been found. The former is illustrated by the amino-terminus heterogeneity of yeast invertase (11,12), whereas the latter is illustrated by the carboxyl-terminus heterogeneity of fungal glucoamylase (13). A secondary heterogeneity can also arise from proteolytic processing after secretion. This has been directly demonstrated in the case of Phanerochaete chrysosporium (14). We have also observed extracellular degradation in late stages of a S. commune culture that is only consistent with a protease secretion (3). In addition, we have observed an apparent specific cleavage of the βGase, but as yet have no information on which end of the molecule this cleavage might have occurred.

Glycosylation - Secreted eucaryotic proteins are often glycosylated during the secretion process, and this is true of the fungal cellulases as well (9,10). We were able to incorporate ^{3}H-mannose into the secreted proteins of S. commune; most of this label went into the βGase and CMCase, and to a lesser extent into the Av. Consistent with this finding, ^{35}S-labelled secreted proteins bound to concanavalin A(conA)-agarose, and this binding was largely blocked in the presence of 500 mM α-methyl-D-mannoside (MeMan) (Fig. 3). The total extracellular protein was passed through a conA-agarose column, and separated into three fractions. These fractions were that unbound, that eluted by 10 mM α-methyl-D-glucoside (MeGlc), and that eluted by 500 mM MeMan. Roughly, these fractions imply that the proteins were unglycosylated, glycosylated but having a "complex" structure with little available mannose (16), and glycosylated with mannose available for tight conA binding. As expected, the three fractions contained glycosylated CMCase of increasing molecular size.

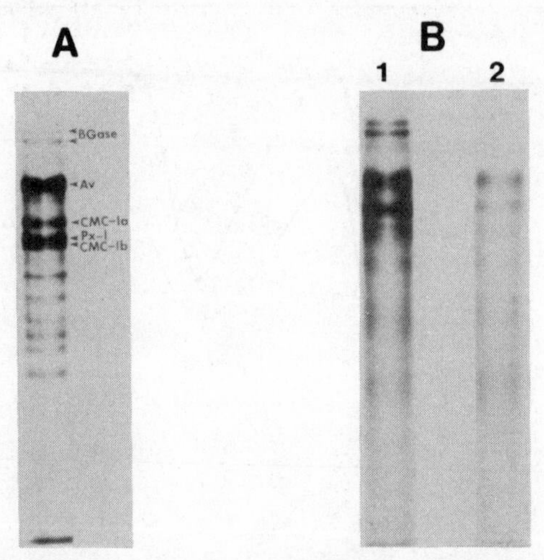

Fig. 3 Glycosylation of S. commune cellulases and the resulting heterogeneity. All samples were electrophoresed on SDS-polyacrylamide gels, treated with fluor, and autoradiographed. The enzymes were labelled with ^{35}S (3).
A. Secretion of enzymes on transfer from glucose to cellulose medium (3).
B. Binding of labelled cellulases to conA (1 1), or to conA in the presence of 500 mM MeMan (1 2).

Heterogeneity of transcripts - We have found that at least two transcripts, coding for two products closely related in molecular weight, are present for each of the cellulases. The anti-CMCase precipitated translation product of mRNA from a 3-day culture is shown in Fig. 4.

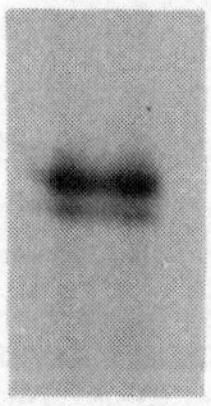

Fig. 4 Multiplicity in CMCase transcripts. The anti-CMCase precipitate was electrophoresed on SDS-polyacrylamide gel, impregnated with fluor, dried, and autoradiographed.

As a result of the above experiments, we have come to the conclusion that much of the observed heterogeneity of the secreted cellulases comes about intracellularly as a result of transcriptional and/or mRNA processing, and glycosylation differences. These variations likely explain part of the secreted cellulase heterogeneity in other cellulase secreting fungi as well as S. commune. Additional heterogeneity also arises from extracellular proteolytic cleavage, for many of the fungi also secrete proteases in addition to their cellulolytic enzymes (3,14).

CLONING OF CELLULOLYTIC GENES

Cloning of vectors and construction of banks - We have constructed several genomic banks of S. commune DNA in bacteriophage-λ derived vectors, as described in Seligy et al. (6). These vectors included λCharon phage 3A, 4A (EcoRI cloning site) and λCharon phage 30 (EcoRI, HindIII, and BamHI cloning sites available). We have more recently used a λvector especially designed to express cloned DNA fragments at a high level suitable for immunoscreening. This bank has been successfully used to screen for the CMCase gene, and this is described more fully below.

Screening of cellulase cDNA clones - A problem associated with the immuno-screening of eucaryote clones in E. coli has been that the expressed foreign proteins are not always stable, and consequently sufficiently high levels of the cloned product are not built up for the immunoscreening. One of the vectors recently constructed to circumvent this problem is designated λgt11 (17). Some of the properties of the vector and its host E. coli mutant include: fusion of the clone to an almost full length copy of the E. coliβ-galactosidase structural gene to help confer stability against E. coli proteolytic degradation, the use of a host mutant that is protease deficient, and the presence of a temperature sensitive mutant gene for the lac repressor carried in a host plasmid. Induction of the cloned product is finally accomplished by laying a nitrocellulose disc against the plate containing the colonies with phage in lytic growth, the disc having been saturated with the lac inducer isopropyl-thio-β-D-galactoside (IPTG). The protein released by λ-induced cell lysis is adsorbed onto the disc. The procedure is outlined in Fig. 5

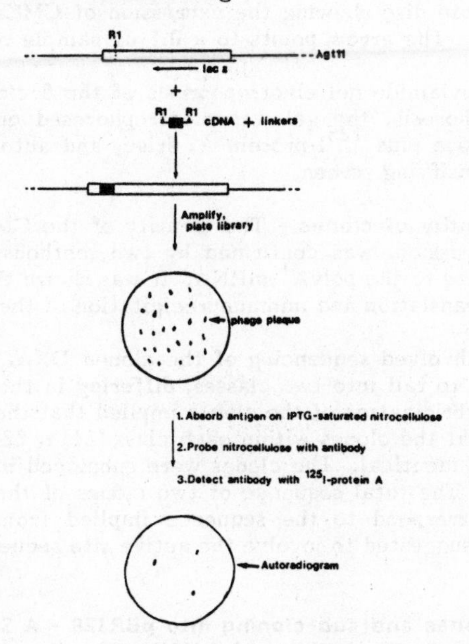

Fig. 5 Screening of λgt11 - S. commune cDNA clones. Adapted from Young and Davis (17).

Selection of CMCase and betaGase clones - Antibodies to the CMCase and βGase were used to select clones from a bank of cDNA in λgt11. The cDNA was prepared from polyA$^+$-mRNA extracted from 3-day cultures of S. commune in a cellulose medium. The selection and purification of the clones are shown in Fig. 6.

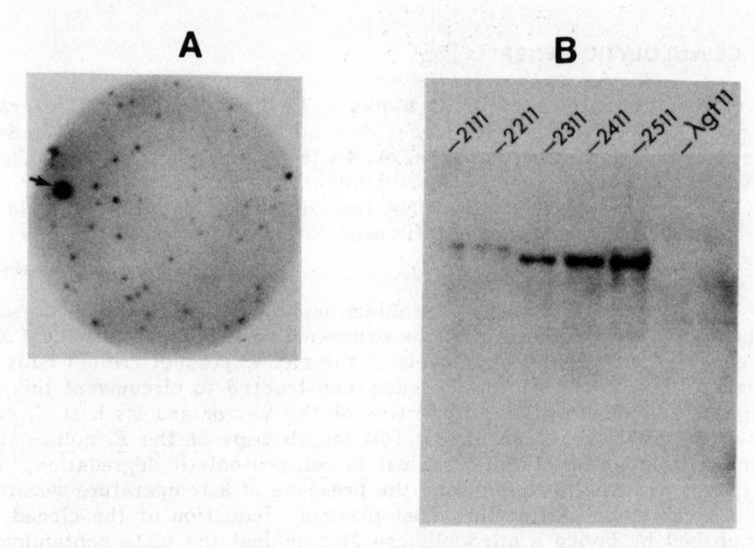

Fig. 6 Isolation of cDNA clones coding for CMCase. A total of 5 clones were independently selected from a bank of S. commune cDNA in λgt11. The procedure followed that shown in Fig. 5.
A. Nitrocellulose disc showing the expression of CMCase antigen by one of the purified clones. The arrow points to a 0.1 ug sample of CMCase spotted on the agar.
B. SDS-polyacrylamide gel electrophoresis of the 5 clones plus a λgt11 control. After electrophoresis, the gel was electrophoresed onto nitrocellulose, probed with anti-CMCase plus ^{125}I-protein A, dried, and autoradiographed in the presence of an intensifying screen.

Confirmation of the identity of clones - The identity of the CMCase clones, selected on the basis of immunoexpression, was confirmed by two methods. In the first method, a clone was used to hybridize to the polyA$^+$-mRNA. It was shown that it selected the mRNA coding for CMCase, by translation and immunoprecipitation of the product.

The second method involved sequencing of the cloned DNA. It can be seen in Fig. 6 that the clones appeared to fall into two classes, differing in the molecular weight of the cloned product. Cross hybridization of the clones implied that the two classes shared some common sequence and that the clones within each class (2111, 2211, and 2311, 2411, 2511) were very similar, if not identical. The clones were subcloned into the sequencing vector M13mp8 and sequenced. The total sequence of two clones of the cDNA have been determined and shown to correspond to the sequence implied from the known amino acid sequence, including that suggested to involve the active site sequence (1).

Selection of genomic clones and sub-cloning into pBR328 - A S. commune genomic DNA library was constructed in λ-phage Charon 3A (18). This library was screened by plaque hybridization (19) for the presence of sequences homologous to ^{32}P-labelled cDNA inserts from λgt11 clones 2111 and 2311 (see Fig. 6B). Five positive clones were isolated; an EcoRI fragment was obtained which was the same for all five clones. This fragment was about 7.1 kbp in length. It was purified by electrophoresis and ligated into the EcoRI site

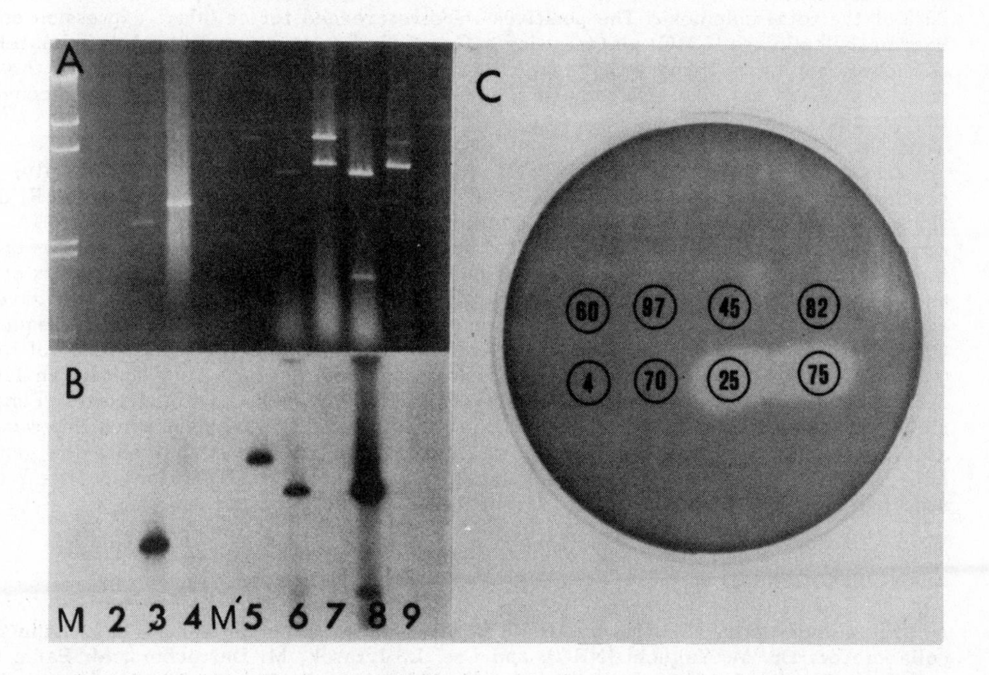

Fig. 7 Expression of truncated S. commune chromosomal CMCase-I gene in E. coli HB101.

A. 1.5% agarose elelectrophoresis of plasmids obtained from subcloning of 7.1 kbp EcoRI genomic fragment in pBR328. The fragment was isolated from 5 separate clones by plaque hybridization (19) using ^{32}P-labelled CMCase-I specific probe (M13mp8-11 and -53 subclones containing inserts from λgt11 clones 2111 and 2311 (see Fig. 6)). HB101 transformants with phenotypes Amp^r, Cm^s, Tet^r were extracted for plasmid and digested with EcoRI. The gel was stained with ethidium bromide and photographed. Tracks M and M' correspond to DNA markers from HindIII digested phage and EcoRI digested pBR328 (colony 70, see C), respectively. Tracks 3-9 correspond to DNA from colonies 4, 60, 97, 25, 45, 75 and 82 (see C).

B. DNA-DNA hybridization analysis of plasmid DNA in A, using ^{32}P-labelled CMCase-I cDNA (mp8-11 and -53) as probe. DNA blotting, labelling, and hybridization were carried out as previously described (21). DNA from tracks 3, 4, 5-9 all carry some of the 7.1 kbp genomic sequence (V.L. Seligy, unpublished data); plasmids from colonies 25 and 75 contain similar sized EcoRI inserts of about 1.4 kbp.

C. Congo Red plate assay for CMCase activity of HB101 transformants. Colonies carrying plasmids described in A and B were grown 16 hr at 30° on 1% CMC, 0.5% yeast extract, 0.7% bactopeptone, 30 ug/ml ampicillin. Clones expressing CMCase show as colonies with unstained zones (clones 25 and 75). Control HB101 (colony 70) contains only pBR328.

of pBR328. Presence of the fragment disrupts the chloramphenicol (Cm) gene function of the plasmid, and the resulting transformants are chloramphenicol sensitive (Cm^S). Plasmids were transformed into the HB101 cells, and selected for Ap^r, Cm^s, and Tet^r (about 30% of the total colonies). The positives were rescreened for cellulase expression on carboxymethylcellulose (CMC) plates, using a Congo Red stain assay (20). Approximately 1% of the colonies were found to have halos on Congo Red stained plates, indicating the presence of CMCase activity. On replating, most of the colonies were found to have converted to Ap^r, Cm^s, and Tet^s.

A selection of the clones (expressing and non-expressing CMCase activity) were checked for the presence of the 7.1 kbp EcoRI fragment insert in pBR328 by EcoRI digestion, electrophoresis, and DNA-DNA hybridization with the CMC-I cloned cDNA. The results indicated most of the plasmids had undergone rearrangements and deletions of both the genomic fragment and the 4.9 kbp pBR328 vector. Fig. 7A-C shows the results of such an analysis carried out on a mini preparation of Tet^s plasmids. No plasmid gave the expected fragment sizes, signifying the presence of intact genomic and pBR328 sequences. Only clones 4, 97, 25, and 75 were found to carry the 360 b.p. of cDNA sequence of CMC-I (Fig. 7A and B, tracks 3, 5, 6, and 8). Plasmids 25 and 75, which both contain an 1.4 kbp EcoRI fragment, were positive when assayed on the Congo Red stained plates (Fig. 7C). Since all plasmids showed homology with the 7.1 kbp genomic fragment when this was used as a genomic probe (data not shown), the simplest explanation is that both the genomic DNA and vector have undergone rearrangement and deletion. Deletions of the 7.1 kbp sequence most likely are required to facilitate expression in E. coli.

ACKNOWLEDGEMENTS

The authors thank L.H. Huang, J.-R. Barbier, and M. Dove for technical assistance and collaborators Dr. M. Yaguchi (NRC), and Drs. L. Jurasek, M. Desrochers, M. Paice (Pulp and Paper Research Institute of Canada) for helpful discussions during the course of this work.

This is NRCC Publication number 24149.

REFERENCES

1. Paice, M.G., M. Desrochers, D. Rho, L. Jurasek, C. Roy, C.F. Rollin, E. De Miguel, and M. Yaguchi. (1984), Biotechnology 2, 535-539.

2. Desrochers, M., L. Jurasek, and M.G. Paice. (1981). Appl. Environ. Microbiol. 41, 222-228.

3. Willick, G.E., R. Morosoli, V.L. Seligy, M. Yaguchi, and M. Desrochers. (1984). J. Bact. 159, 294-299.

4. James, A.P., V.L. Seligy, D.Y. Thomas, M. Yaguchi, G.E. Willick, R. Morosoli, and M. Desrochers (1981), Proc. Third Bioenergy R&D Seminar, Ottawa, 135-139.

5. James, A.P., V.L. Seligy, D.Y. Thomas, M. Yaguchi, G.E. Willick, R. Morosoli, W. Crosby, M. Desrochers, and L. Jurasek. (1982), Proc. Fourth Bioenergy R&D Seminar, Winnipeg, 425-429.

6. Seligy, V.L., J.-R. Barbier, K.D. Dimock, M.J. Dove, F. Moranelli, R. Morosoli, G.E. Willick, and M. Yaguchi. (1983), in Gene Expression in Yeast. Proc. ALKO Yeast Symposium, Foundation for Biotechnical and Industrial Fermentation Research (1983), 167-195.

7. Seligy, V.L., M.J. Dove, C. Roy, M. Yaguchi, G.E. Willick, J.R. Barbier, L. Huang, and F. Moranelli. (1984), in Proc. Fifth Canadian Bioenergy R&D Seminar, Ottawa, S. Hasnain (Ed.), Elsevier (New York), 577-581.

8. Eriksson, K.-E. (1981), in Trends in Biology of Fermentations. A. Hollaender (Ed.), Plenum, New York, pp. 19-31.

9. Gum, E.K. Jr, and R.D. Brown, Jr. (1977), Biochim. Biophys. Acta 492, 225-231.

10. Berghem, L.E.R. and L.G. Pettersson. (1973), Eur. J. Biochem. 37, 21-30.

11. Perlman, D. and H.O. Halvorson. (1981), Cell 25, 525-536.

12. Carlson, M., R. Taussig, S. Kustu, and D. Botstein. (1983), Mol. Cell. Biol. 3, 439-447.

13. Boel, E., I. Hjort, B. Svensson, F. Norris, K.E. Norris, and N.P. Fiil. (1984), EMBO Journal 3, 1097-1102.

14. Eriksson, K.-E. and B. Pettersson. (1982), Eur. J. Biochem. 124, 635-642.

15. Coughlan, M.P., P.J. Considine, A.P. Moloney, and A.J. McHale. (1982), TAPPI Proceedings, Research and Development Conference, Asheville, North Carolina, pp. 75-80.

16. Cummings, R.D. and S. Kornfeld. (1982), J. Biol. Chem. 257, 11235-11240.

17. Young, R.A. and R.W. Davis. (1983), Science 222, 778-782.

18. Blattner, F.R., B.G. Williams, A.E. Blechel, K. Derriston-Thompson, H.E. Faber, L. Furlong, D.J. Grunwald, D.O. Kiefer, D.D. Moore, E.L. Sheldon, and O. Smithies. (1977), Science 196, 161-169.

19. Benton, W.D. and R.W. Davis. (1977), Science 196, 180-183.

20. Beguin, P. (1983), Anal. Biochem. 131, 333-336.

21. Seligy, V.L., D.Y. Thomas, and B.L. Miki. (1980), Nuc. Ac. Res. 8, 3371-3393.

Section 5

Other Biomass Conversion Technologies

CALCIUM MAGNESIUM ACETATE FROM BIOMASS

H.R. Bungay* and L.R. Hudson

New York State Energy Research and Development Authority
Rockefeller Plaza, Albany, N.Y. 12223

ABSTRACT

A switch from salt or calcium chloride to calcium-magnesium acetate for de-icing roads and highways would greatly decrease corrosion of automobiles, bridges, and highways, and deterioration of underground cables and pipes. Low-cost dolomitic lime used to neutralize acetic acid determines the proportions of calcium and magnesium. Cost will be critical, hence, the use of inexpensive fermentation substrates deserves prime consideration. Possible processes include the fermentation of cellulose, anaerobic digestion of biomass, and a non-fermentative route using alkaline degradation of carbohydrates.

INTRODUCTION

The sodium chloride and calcium chloride used for de-icing roads in the Northern U.S. greatly accelerate corrosion of automobiles, damage concrete and asphalt, ruin bridges, and have many other drawbacks, such as adverse effects on underground cables. There are also many subtle costs of de-icing with corrosive salts; for example, gasoline wasted because of slowed traffic during repairs to bridges and roads, the cost of transporting materials for repair, and labour and materials for replacement. In New York State alone, the bad effects of de-icing cost more than $500 million per year. About one million tons of salt are used each year in the State at a cost of $100 million. The salts that are flushed away sometimes reach natural waters, lowering quality and affecting taste and utility. Of the many substitutes that have been proposed, a mixture of calcium acetate and magnesium acetate has great promise because of affordable cost, reasonable effectiveness for de-icing, and mild corrosive properties. Inexpensive, locally available dolomitic lime (a mixture of mostly calcium and magnesium compounds) determines the cations associated with acetate. A typical dolomitic lime has the following weight percentages: 47.5 CaO, 34.3 MgO, 1.8 $CaCO_3$.

Preliminary field testing of calcium-magnesium acetate (CMA) began last winter in Michigan and Washington, but results have not yet been reported. Studies are underway to determine the environmental impacts of large-scale use of CMA. There will be some increase in B.O.D. when CMA is flushed into natural waters, but oxygen may not be seriously depleted because this pollution occurs when temperatures are low and biological systems function slowly. Damage to soils is not likely from CMA because organic matter is desirable, and calcium and magnesium ions are already present in large amounts. Oxidation of CMA by soil microorganisms will generate carbonate species that are harmless.

The New York State Energy Research and Development Authority (NYSERDA) is interested in CMA as a salt replacement for the following reasons:

1. New York is well endowed with low grade hardwoods, and is therefore capable of becoming a large producer of wood sugars (through hydrolysis). Well-known fermentations of glucose from cellulose lead to ethanol, glycerol, 2,3-butanediol, acetone, butanol, and many other chemicals that could relieve dependence on fossil feedstocks. However, good applications for sugars from hemicellulose are harder to find; CMA development may hasten the commercialization of fuels and petroleum-displacing chemicals in New York and the growth of a new industry based on indigenous resources.

2. Repairs to corroded underground electric feeder systems cost ratepayers in New York over $100 million a year. By eliminating the damage caused by salt, a significant contribution to lowering the cost of electric power in the State may be achieved.

Adequate supplies of CMA are not available, and samples for testing on roads have been made by reacting acetic acid with dolomitic lime followed by drying. Cost is very important because CMA is not as effective as salt or calcium chloride in lowering the freezing point, although CMA is more soluble in water. About 50 to 100 per cent more CMA must be applied, resulting in greater transportation and labour costs. An early target price is 18.5 cents per pound, but 10 cents would be more desirable for rapid acceptance. Quality is not particularly important for de-icing, and a fermentation route to crude acetic acid seems very attractive. Large-scale evaluation of the environmental impacts of CMA should use material, including the impurities, typical of a practical production process.

New York State's requirements for CMA would be about 1.7 million tons per year. If produced from hydrolyzed hemicellulose sugars, about 6.8 million tons of hardwood chips would be needed, or about half the excess growth on commercial forest land in the State. Approximately 500 million gallons of ethanol could be a co-product based on cellulose fraction. Inexpensive sugars are also available from the waste streams of conventional paper pulping. These sugars could be obtained prior to pulping, with fewer impurities, by dilute acid extraction of the wood chips.

A comparison of ethanol fermentation and CMA fermentation is interesting. The yields are as follows:

1 kg of glucose yields 0.48 kg of ethanol
1 kg of glucose yields 0.85 kg of acetic acid or 1 kg of CMA
1 kg of xylose yields 0.85 kg of acetic acid or 1 kg of CMA

Most of the non-organic carbon leaves the system as carbon dioxide that could be collected and sold for additional revenue. The higher weight yield would indicate that CMA could be a lower-cost product than ethanol. Furthermore, CMA is produced through a relatively easier fermentation process. While the best organisms for ethanol ferment only a few sugars, several known organisms will use a variety of sugars to produce acetic acid. Most organisms have a reasonably good tolerance to acetic acid although a low pH can be inhibitory. However, the continued addition of dolomitic lime to control the pH would solve the pH problem.

Sugars obtained from hemicellulose should cost less than that of other fermentable carbohydrates because the potential sources of hemicellulose are vast and there are few competing uses for pentoses. Kirk et al. (1983) estimated that the pulp and paper industry produces about 1.2 metric tons per year of waste sugars from sulfite mills, 1.5 t/y of cellulosic material from Kraft mills, and 150,000 t/y of sugars from hardboard and insulation board plants. Biomass pretreatment could increase these totals. These sugars are currently held in low regard because of contamination by furfural, resins, and toxic organic compounds. Although they may be used in cattle feeding, biomass syrups containing pentoses may command a price lower than that of molasses. Nevertheless, organisms that can tolerate the presence of these impurities may be selected for effective fermentation to CMA, or a simple clean-up of these sugars prior to fermentation should be possible.

PROPOSED PROCESSES

The production of acetic acid by fermentation of corn syrup uses Clostridium thermoaceticum in research studies at the University of Georgia (Anon., 1984). This organism also ferments xylose and probably would make good use of the mixed sugars from the hydrolysis of hemicellulose.

For several years, columns packed with wood shavings, to which acetobacter are attached, have been used to produce vinegar. A similar system might be effective for producing CMA. Options that deserve attention are the use of limestone as the packing material or the use of a feed stream with suspended lime.

The acidogenic phase of anaerobic digestion often gets ahead of the methanogenic phase and results in a "sour" digester. This can happen when the digester becomes overloaded, and the concentration of mixed organic acids exceeds 0.5 per cent. Although this is too low for a commercial CMA process, an analogous technology may produce higher acid concentrations if the feed stream is sugar or partially purified cellulose. The advantage would be avoidance of aseptic conditions, although the product would contain the salts of several organic acids. However, acetic acid should predominate.

Although the sugars from hemicellulose are probably the cheapest substrate for making CMA, cellulose from biomass refining may also be used. The production of organic acids from sugars derived from cellulosic biomass and the preliminary cost estimates have been reported recently (Clausen and Gaddy, 1984; Jones et al., 1984). Cellulose from a biomass pretreatment using organic solvents would not be a suitable fermentation substrate because its long fibers have a much higher value for manufacturing paper or chemical derivatives. Cellulose obtained from the steam-explosion process, however, has a low degree of polymerization and a fiber length unsuitable for paper making or chemical derivatives. The current plans of several companies are based on the conversion of such cellulose to fuel-grade ethanol. As discussed before, it may be more profitable to make CMA instead of ethanol. Possible routes include enzymatic hydrolysis of cellulose followed by fermentation of the glucose to acetic acid, or direct fermentation of the cellulose. The direct fermentation of lignocellulosic biomass (Avgerinos and Wang, 1981) requires two organisms - one to hydrolyze both cellulose and hemicellulose and the other to ferment the mixed sugars, since the first organism ferments only glucose. With partially solubilized cellulose, only one organism may be required. In addition, it may be easier to improve tolerance to CMA than to ethanol.

A serious problem in developing direct fermentation of cellulose-to-ethanol technology has been to suppress the formation of acetate and lactate by the selection of better mutant cultures. Some of the inferior strains for ethanol production may have good potential for producing acetate.

Non-fermentative routes from carbohydrates to acids should not be overlooked. Cooking carbohydrates with alkali at moderately elevated temperatures produces salts of mixed acids. Cellulose, cooked with an aqueous alkali, gives predominantly lactate plus lesser amounts of formate, acetate, and glycols (Krochta et al., 1984). The South Dakota Department of Transportation has a recent patent for cooking sawdust or wood wastes with dolomitic lime to produce calcium and magnesium salts of lactate, acetate, and glycollate (Anon., 1984). It is claimed that this mixture may be better in some respects than CMA for de-icing.

Testing of various sugars cooked with sodium hydroxide or calcium hydroxide showed that sucrose gave about 80 per cent weight yield of salts, while no other sugar gave more than half this amount (Allen et al., 1984). If these reactions were better understood and if better catalysts were to be found, the non-biological routes to salts or organic acids could have excellent commercial prospects. Lactate is valuable but has limited markets, so increasing the proportions of acetate deserves high priority for research. It should be noted that inexpensive CMA would be a practical intermediate for making acetone by an ancient pyrolysis process.

CONCLUSION

An exciting new opportunity for using biomass is the manufacture of CMA. If lignin can be sold for a good price for specialty applications, or even for bulk uses, such as supplementing asphalt, the combined revenues with CMA look very attractive. This should not delay the ultimate commercialization of fuel-grade ethanol production because the processes (CMA and ethanol) have much in common. The availability of large amounts of untreated biomass and partially pretreated cellulose should stimulate research and development of ethanol processes. A factory may produce CMA seasonally for the de-icing market and produce ethanol the rest of the year so as to minimize storage costs.

REFERENCES

Allen, B.R., W.J. Huffman, and E.S Lipinsky. "Thermochemical Production of Lactic Acid". Abstracts of Papers of 6th Symposium on Biotechnology for Fuels and Chemicals, Gatlinburg, TN. 1984.

Anon., "CMA Road De-icer Studies Progress, Alternatives Found". Biomass Digest, 6:7 1984.

Avgerinos, G.C., and D.I.C. Wang. "Direct Microbiological Conversion of Cellulose to Ethanol". Ann. Rept. on Ferm. Proc. 4: 165-191 1980.

Clausen, E.C., and J.L. Gaddy. "The Production of Organic Acids from Biomass in Continuous Fermentation Reactors". Paper presented at A.I.Ch.E. Summer National Meeting, Philadelphia. (1984).

Jones, J.L., C.W. Marynowski, D. Tuse, and R.L. Boughton. "Fermentation for the Production of an Acetate De-icing Salt". Paper presented at A.I.Ch.E. Summer National Meeting, Philadelphia. (1984).

Kirk, T.K., T.W. Jeffries, and G.F. Leatham. "Biotechnology: Applications and Implications for the Pulp and Paper Industry". TAPPI 66: 45-51 1983.

Krochta, J.M., J.S. Hudson, and C.W. Drake. "Alkaline Degradation of Cellulose to Organic Acids". Abstracts of Papers of 6th Symposium on Biotechnology for Fuels and Chemicals, Gatlinburg, TN. 1984.

PRODUCTION OF USEFUL METABOLITES BY PLANT TISSUE CULTURE

Masanaru Misawa

Allelix Inc.
6850 Goreway Drive
Mississauga, Ontario Canada

Since the end of the 1980's, many researchers have studied the production of cell mass and useful metabolites using plant tissue culture technology. Higher plants contain a variety of useful substances which have been useful as medicines, food additives, perfumes etc. Thus the decrease of plant resources, increase in the labour cost, and other problems in obtaining these substances from natural plants have provided motoviation for using plant cell culture. The method is not affected by changes in environmental conditions such as climate, therefore, production is available in any place, in any season.

In 1959, Tulecke and Nickell described the first large scale culture system for plant cells. Since then, Mandels, Street, Staba, Martin and other research groups have studied large scale cultivation using a variety of fermentors. These studies have stimulated more recent studies on the industrial application of plant tissue cultures in West Germany, Japan and many other countries. For example, in 1982, the 5th International Congress of Plant Tissue Culture was held in Japan and approximately 70 out of 370 (the largest number ever held) papers presented there related to production of cell mass and of secondary metabolites. A Japanese company recently started the production of shikonin derivatives in industrial scale for the pigment of cosmetics.

In spite of remarkable advances in plant tissue culture technology, however, the production cost of metabolites is still high for the production of food stuffs, food additives or pharmaceuticals which can be more easily produced by chemical synthesis or by fermentation. Therefore, it is very important to choose what kind of products we should make through tissue culture technology.

PRODUCTION OF PLANT BIOMASS

For the purpose of producing food stuff, cells of bean, lettuce, carrot etc., were cultivated in the fermentors in the 1960's by Mandels et al. It would seem to be attractive if vegetables could be produced in tanks at any time, but this is not an economical way because the cost of these vegetables is too low to use this technology.

Tobacco Cells

Since 1970, studies on tobacco cell cultures have been carried out by the research group in the Japan Tobacco and Salt Public Corporation to obtain a uniform quality of tobacco raw materials independent from climatic and geographical factors. They reported semicontinuous culture of Nicotiana tabacum strain with a 30 L jar fermentor for 4 days in 1972 and then they succeeded in continuous culture with a 1500 L fermentor in 1976. A higher level of inorganic nitrogen compounds in the medium was found to be favourable for rapid growth of the cells, but it gave undesirable results in quality. In order to obtain the cells of a relatively low nitrogen content with high productivity as possible, a two stage, two stream culture has been carried out by the group (Fig. 1). In this system, fresh standard Murashige Skoog's medium with a nine tenth nitrogen source and 3 times phosphate at a dilution rate of 0.54 and the inflow is balanced by the outflow of the corresponding volume of the culture which is transferred to Tank II. Tank II was supplied with another MS medium with one sixth of the normal nitrogen source and 3 times stronger phosphate. They obtained the cells continuously at the rate of 6.9 g d.w. cells/L/day. Using a 20 kL fermentor, they could cultivate cells continuously for 66 days and the cell level was 16.5 g/L with a production rate of 5.85 g/L/day.

Ginseng

Panax ginseng is a perennial indigenous to Eastern Asia and cultivated in China, Korea and Japan. Ginseng root, so-called "Ginseng radix" has been widely used as a tonic and natural medicine in oriental countries. Since cultivation of P. ginseng in the field requires four to seven years and since it is impossible to plant consecutively for 20 to 50 years, Furuya et al. isolated a cell line of this plant having high yields of saponins and cell mass and cultivated this in 30 L jar fermentors. After 28 days cultivation, the highest level of cell mass, 17 g per liter was obtained. The cells contained about 50 mg of total saponins per liter of the medium. The industrial application is being considered by a Japanese company.

Bupleuri radix

As a crude drug, the dried root of Bupleurum falcatum, named "Bupleuri radix" has been applied in oriental countries as an antipyretic, a tonic and anodyne. According to a patent by Tomita et al., the cells of the plant regenerated roots and primordia in a tank and accumulated about 19 mg of saponins per gram of the cells which was the same level as had accumulated in the natural plants.

The two examples relate to production of the crude drugs which are already in the market using plant tissue cultures. The market of the crude drugs has been expanding and therefore this sort of approach is expected because of their high added value.

Plant Virus Inhibitors

The author and his group at Kyowa Hakko Co. Ltd. have worked on production of anit plant virus substances by Phytolacca americana and Agrostemma githago suspension cultures. The active principles were shown to be proteinaceous compounds and were very active against TMV and CMV (Table 1). Since the purified compound is not suitable for the field in practice because of the high purification cost, the cells of both strains were treated with a homogenizer for 5 minutes and the supernatant obtained by centrifugation was diluted to an appropriate concentration with water. The solution was applied to leaves of various plants to inhibit virus infections. The level of the active principles increased in response to the increase of cell mass, so many experiments have been done in order to increase the cell mass.

PRODUCTION OF SECONDARY METABOLITES

As I described above, production of cell mass by tissue culture technique is useful if the cells have biological activities such as Panax ginseng and Phytolacca americana. However, many research groups have been investigating production of secondary metabolites using plant tissue culture rather than that of biomass, and they have been trying to increase producing efficiency per cell in order to diminish the high production cost by several approaches as shown in Table 2.

Optimization of Environmental Conditions

This is one of the most common approaches in which a variety of chemical and physical factors affecting cultivation have been tested extensively in many kinds of plant cells. These factors include components of the medium, phytohormones, pH, temperature, aeration, agitation, etc.

Among many compnents in the medium, sucrose is the most popular carbon source for plant tissue culture and the level of the sugar also affects the productivity. For example, higher levels of sucrose stimulated the level of anti-plant virus substances by Agrostemma githago cultures.

The kind and level of inorganic nitrogen sources in the medium are also important factors for growth and production. As we have shown, the level of L-glutamine accumulated in the cells of Symphytum officinale increased to about 20% of dry weight in the medium containing high levels of nitrogen compounds.

Phosphates often stimulate cell growth but sometimes decrease the level of alkaloids in the cells. Sasse et al. in West Germany found that the activity of enzymes in Catharanthus roseus diverting the corresponding primary precursors into secondary pathways such as phenylalanine ammonialyase, was reduced in the medium containing high levels of phosphates.

Ca^{++} and other major or minor elements in basal medium have an effect on the growth and the productivity in some cases. We found that the amount of Agrostemma githago cells adhering to the wall of the fermentor was decreased markedly and the cells were easily removed from the inside wall of a fermentor to the medium by reducing the level of $CaCl_2aq$ in MS medium. This is particularly advantageous for large scale cultivation.

Among a number of components in the medium, phytohormones have the most remarkable effects in growth and productivity. In general, the increase of auxin levels, such as 2,4-D in the medium stimulates the differentiating of the cells and consequently diminishes the level of secondary metabolites. However, it is not always true since production of L-DOPA by Mucuna pruriens, of Ubiquinone-10 by N. tabacum and of diosgenin by Dioscorea deltoidea were stimulated by high levels of 2,4-D. N-(2-Chloro-4-pyridyl)-N'-phenylurea, so-called 4-PU-30 is a new type of compound, by Tripterygium wilfordii suspension cultures.

Recently, the Mitsui Petrochemical Co. in Japan established a method in industry to produce shikonin derivates using Lithosperum erythrozhizone cell cultures. The level of shikonin derivatives in a modified White medium was much higher than that in the original medium. Using a 750 L fermentor, about 1.5 grams of shikonin derivatives per liter were obtained and its level corresponded to 14% dry weight cells. This is the first case in industrial application of secondary metabolite production by plant cell culture. The product is being produced as a pigment for a lip stick.

As seen in papers and reviews, physical conditions for cultivation such as aeration, agitation and so on are also very important factors, particularly in large scale fermentation. In anthraquinone production by Morrinda citrifolia, Wagner et al. suggested an air-lift type fermentor was the most suitable for plant cell cultures. In the case of biotransformation of β-methyldigitoxin to β-methyldigoxin by immobilized cells of Digitalis lanata, no difference in productivity was recognized between Erlenmyer flasks and bubble columns.

Addition of Precursors and Biotransformation

Addition of appropriate precursors or related compounds to the culture media sometimes stimulates the producing ability. Amino acids have often been added to the media for production of tropane alkaloids and indole alkaloids and some stimulative effects were observed. It is true that some amino acids are precursors of various alkaloids but generally the biosynthetic steps from amino acids to alkaloids are so complicated that I doubt whether the amino acids added were incorporated into the alkaloids in cell culture. Perhaps, they affected not only the biosynthesis of the alkaloid directly as precursors but also other various metabolic pathways in the cells, activating synthesis indirectly.

We found that the level of tripdiolide was increased by addition of farnesol. Farnesol is a dephosphorylated compound of farnesyl pyrophosphate which is an intermediate in the biosynthesis of terpenoids.

Biotransformation of β-methyldigitoxin toβ-methyldigoxin using Digitalis lanata cells has been a possibility since 1974 because of the promising study by Reinhard and Alfermann. To reduce the production cost, which is very important in industrial applications, they examined the use of immobilized cells and semicontinuous cultures. The maximum level of the product was 700 mg/L.

In order to produce prodrug, glucosylation of salicylic acid using Mollotus japonicus cells was studied and its conversion yield was higher than 90%. The maximum level of salicylic acid-o-glucoside obtained was 0.9 g/L. It is of interest that the glucoside showed as potent an analgesic activity as salicylic acid, while its effect was more rapid and more long-lived than that of salicylic acid in mice.

Selection of High Producing Cell Lines

The physiological characteristics of individual cells are not always uniform. For example, it is observed that a pigment producing cell aggregate consists of producing cells and non-producing cells. In 1976, Zenk and his colleagues obtained cell lines of C. roseus which could produce high levels of ajmalicine and serpentine as determined by radioimmunoassay. Since their excellent results, a number of researchers have examined the cell cloning method because this is the most helpful way of increasing the levels of metabolites present.

Table 3 shows several examples of cell cloning carried out in Japan. Most of them are related to production of pigments such as anthocyanines since selection was easy because of their visibility.

Berberine is useful as an intestinal antiseptic in the Orient. The selected strain of Coptis japonica produced 1.2 g of berberine per liter of the medium. A group in the Japan Tobacco and Salt Public Corporation isolated a number of strains producing high levels of ubiquinone-10 from tobacco cell cultures. After the 13th recloning, a strain was selected from 4,000 cell clones, and the level was 5.2 mg/g dry weight cells which corresponds to 180 times the amount of the parent in the mother plants.

The cell cloning method is undoubtedly a very useful technique to increase the level of secondary metabolites, and it should be applied on many other cultures. However, it is not obvious why cells contain both high- and low-yielding strains and only few papers concerned with the mechanism have appeared.

Berline et al. compared the highest cinnamoyl putrescines producing p-fluorophenylalanine resistant strain of N. tabacum cv. Xanthi with the low producing strain in five enzymes of the biosynthetic pathway. As a result, activities of these enzymes were found to be 3 to 10 times higher in the high producing strain. This sort of study will be very useful to induce mutants which are able to overproduce useful plant metabolites.

Induction of Variants

In industrial fermentation technology, induction of genetic mutant strains of microorganisms is ubiquitous, however, vegetative plant cells are normally diploid and it is much more difficult to obtain genetic mutants. Furthermore, biosynthesis pathways of secondary metabolites and their regulation mechanisms in higher plants are mostly obscure, therefore, it's also difficult to decide what kind of mutants should be induced in order to increase the producing ability.

As I mentioned before, Berline and his colleagues reported that a p-fluorophenylalanine resistant strain of N. tabacum produced 6 to 10 times higher levels of cinnamoyl putrescine than that of the parent strain. They also found that a parent strain of C. roseus produced catharanthine only in the production medium but its tryptophan analogue resistant mutant accumulated the same alkaloid even in the growth medium.

Increase of metabolite levels using regulatory mutants is theoretically possible and selection of suitable analogues for this purpose could be an important factor in order to produce a variety of products.

Application of Immobilized Cells

Since the growth of most plants is generally lower than that of bacteria, culture must proceed for longer. Use of immobilized cells instead of free cells is sometimes advantageous. Brodelius et al. used Morinda citrifolia immobilized cells by entrapment in alginate to produce anthraquinone.

It was found that the immobilized cells had an increased synthesis of anthraquinones as compared to freely suspended cells under the same conditions. They also recognized the synthesis of indole alkaloids in 10% yield from tryptamine and secolaganine using immobilized C. roseus. Although the products are normally stored in the vacuoles of the cells, addition of a trace amount of chloroform in the aqueous phase altered cell permeability and consequently the products were excreted extracellularly.

The biotransformation from β-methlydigitoxin toβ-methyldigoxin using alginate entrapped Digitalis lanata was also studied by Alferman et al. The reaction was carried out at a constant rate for at least 150 days after a lag phase of about 20 days.

Recently, an efficient biotransformation of codeinone to codeine in 70% yield was reported using immobilized Papaver somniferum cells. Most of codeine was excreted extracellularly.

Morphological Differentiation

There are large numbers of reports describing the loss of secondary metabolite productivity in undifferentiated cells. Therefore, studies have been made using differentiated cells which have roots, shoots or other organs. But it is not clear to what extent secondary metabolism depends on the development of specific structures, and it is known whether these two processes are genetically and/or physiologically linked.

It was reported by Hiraoka et al. that stems, shoots and roots were differentiated by turn in Datura meteloides cells when the callus was successively transferred from a medium with auxin to that without auxin and tropane alkaloid level increased. A similar phenomena was also found in rotenone formation using Derris elliptica and morphine alkaloid production using Papaver somniferum.

In the case of production of potent anti-tumor alkaloids, cephalotaxine esters, using Cephalotaxus harringtonia, the differentiated tissues produced them in higher level.

Thus, it seems that morphological differentiation is necessary to obtain higher yields of secondary metabolites in many cases, but this is not always desirable in large scale cultivation, because generally the culture period for differentiated tissue is rather longer than that of undifferentiated cells. It is obvious that a shorter culture period is required to avoid microbial contamination and to lower the producing cost.

In conclusion, with the advance of biotechnology, research on plant tissue culture is becoming more active not only in academic institutions but also in industry. From an economical point of view, there are at least three applications of this field, namely plant breeding, useful metabolite production and micropropagation.

Except for micropropagation through tissue culture, it has been thought that the two other objectives take too long a time to contribute to industry. However, as seen in shikonin derivative production, the industrial application of this technology to manufacture metabolites seems likely to be realized in the near future although there are still some problems.

In order to overcome these problems, it is obvious that more extensive fundamental research will be needed collaborating with a number of researchers in other scientific fields. Furthermore, it should be noted that the selection of the most suitable products for plant cell culture is very important as I have emphasized repeatedly.

Table 1 Inhibition of Tobacco-Mosaic-Virus Infection by Various Virus Inhibitors

Inhibitors	Assay Conc. (μg/mL)	Inhibition[1] (%)
Aabomycin A	1.0	0
Blasticidin S	1.0	75
Miharamycin A,B	1.0	27
P. americana, cultured cell[2]	33.0	76
A. githago, cultured cell[2]	1.0	78
	7.0	91

[1] Assayed by a local lesion method with TMV and P. vulgaris
[2] Purified products from cultured cells

Table 2 Approaches to Efficient Production

1. Optimizing Environmental Conditions
2. Addition of Precursors and Biotransformation
3. Selection of High Producing Cell Lines
4. Induction of Variants
5. Application of Immobilized Cells
6. Morphological Differentiation

Table 3 Typical Examples of Cell Cloning Application in Japan

Products	Plants	Factor	Researchers
Anthocyanins	Vitis hybrid	2,3-4	Yamakawa et al.
Anthocyanins	Euphorbia milli	7	Yamamoto et al.
Berberin	Coptis japonica	2	Satoh et al.
Biotin	Lavendula vera	9	Watanabe et al.
Ubiquinone-10	Nicotiana tabacum	15	Matsumoto et al.

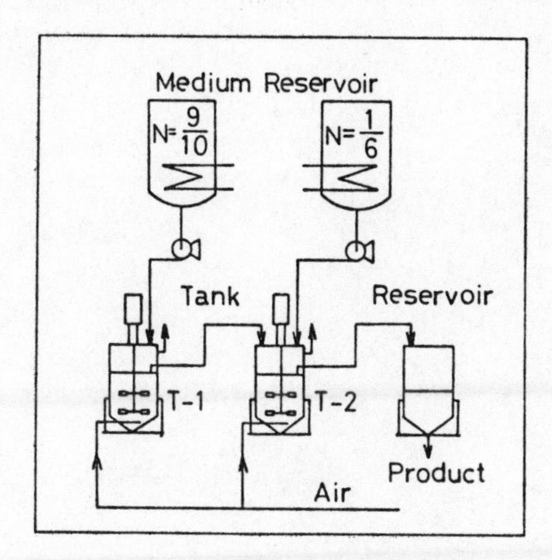

Figure 1. Two stage and two stream culture for tobacco cell culture (M. Noguchi et al.)
T-1: Culture Tank I
T-2: Culture Tank II

PRODUCTION OF LIQUIDS FROM BIOMASS BY CONTINUOUS FLASH PYROLYSIS

Donald S. Scott and Jan Piskorz

Department of Chemical Engineering, University of Waterloo,
Waterloo, Ontario
N2L 3G1 (Canada)

ABSTRACT

A simple, continuous process for flash pyrolysis of biomass has been developed. It employs a fluidized bed at atmospheric pressure, and is capable of giving high liquid yields from lignocellulosic materials. Extensive tests have been carried out in a 15 g/hr bench-scale unit and, over the past year, tests have been done in a larger (2 kg/hr) unit. At optimal conditions, yields of 65% (by wt.) are achieved for production of organic liquids from hardwoods, such as poplar or maple. Various agricultural wastes give liquid yields of 40-65%. Pyrolysis oils produced are suitable for use as low-grade liquid fuels. A medium quality gas and a reactive char are also produced.

Some of the physical and chemical characteristics of the pyrolysis oils are presented. Preliminary economic assessment indicates that the process is attractive even on a small scale (ca. 100 tonnes/day), and heating oil could be produced at competitive prices. In part, this results from a unique operating design for the fluidized bed reactor that allows economical, continuous operation on a small scale.

INTRODUCTION

In three previous publications (1,2,3), we have outlined the development of an atmospheric pressure, flash pyrolysis process utilizing a fluidized bed of solids as heat carrier. The process studies had as a primary objective the determination of conditions for maximum yield of liquids from biomass, particularly forest materials. Results from a bench-scale unit (15 g/hr feed rate) indicated that, at apparent vapour residence times of about 0.5 seconds, organic liquid yields of 60%-70% on a moisture-free basis could be obtained from hardwoods such as aspen-poplar and maple. Organic liquid yields of 40%-60% could be obtained from agricultural wastes, such as wheat straw, corn stover and bagasse. Bench-scale reaction conditions used a biomass particle size of ◘295 μm in a nitrogen atmosphere over a temperature range of 400°C-650°C. Effects of catalyst and lime addition, of particle size, and of other reaction atmospheres (methane and hydrogen) were also reported. In terms of calorific value and carbon to hydrogen ratio, the best quality liquid was obtained at conditions that also gave the maximum liquid yield.

In view of the high yields of organic liquids obtained -- the highest yields yet reported for a pyrolysis conversion process for biomass -- and the reasonable operating conditions, a larger scale continuous process unit was designed and constructed. The liquid yields attained in this unit have been as good as, or better than, those achieved in the bench scale unit. In this report, results are given for three different biomass materials. A description of the process is also given, as well as a preliminary economic analysis.

EXPERIMENTAL

The design of the small pilot plant unit was determined in part by economic requirements for a successful biomass pyrolysis process. The outstanding economic feature of biomass conversion processes is the cost associated with harvesting the raw material. If the biomass is available as a waste at a central facility (e.g. a sawmill), then the supply will be limited. If the biomass must be harvested, this can only be done economically over a restricted radius, possibly 50 km. As a result of this limitation of supply at a site, a biomass conversion process should be economical on a small scale, so that the plant could be located at the source of raw material supply. Products from several such conversion plants could then be shipped readily to a common site for further processing, or to other customers, particularly if the products were liquid.

If a process is to be economical on a small scale, it should operate as simply as possible, and it should require low capital investment. The pyrolysis conditions found to be optimal in the bench scale work required only moderate temperatures, 450°-550°C, atmospheric pressure, and the use of a fluid-bed reactor. Hence, it seemed likely that the required economic criteria could be met.

For the small-scale pilot unit, sawdust or other biomass is air dried and then hammer milled and screened to ■595 μm (■30 mesh) particle size. A twin-screw feeder conveys the sawdust from a hopper into a stream of recycled product gas, which carries the sawdust into the reactor. The feed injection point is within the fluid bed itself. The reactor contains sand as the inert fluidized material, and the fluidizing gas is recycled product gas that has been preheated in the inlet line by controlled electric heaters. In addition, the reactor is wrapped with heating coils allowing extra heat to be added either to the bed of sand or to the freeboard space, as desired.

Pyrolysis products are swept from the reactor to a cyclone, together with all the char formed. The char is separated in the cyclone and the product gas and vapours pass to two condensers in series. These condensers have the pyrolysis gas inside the tubes, which are vertical. Every tube has a clean-out plug at the top and a condensate collection pot at the bottom. The first condenser can be operated at temperatures up to 100°C, while the second condenser uses chilled water at 0°C as a cooling medium. The gas from the condensers then passes through a series of filters to remove tar mist and then a recycle compressor. A regulated gas flow is taken from the compressor discharge to fluidize the reactor and to convey feed into the reactor, while the excess is vented through a gas analyzer and gas meter as product gas.

The method of operation of the fluidized bed reactor is unique to this process. Existing fluid-bed pyrolysis technology normally uses twin fluid beds: the first contains the pyrolysis reaction. The sand and char are then circulated to a second fluid bed where the char is burned and the sand heated. The hot sand is returned to the first bed to supply the required heat of reaction. While this method of pyrolysis is used commercially (e.g. for municipal waste gasifiers), it can have operating problems with transfer lines, and is expensive. In our reactor, the design has been carried out so that during pyrolysis the char formed is blown from the bed and the sand is retained in the bed. By carefully selecting sand size, biomass particle size, bed velocity and reactor configuration this method of operation is highly successful. This "blow-through" method of operation has been used without problems under a variety of conditions in the continuous unit described here. After tests with biomass feeds of several times the sand mass in the bed, it was found that the bed contained a negligible amount of char. The outcome of this special design and mode of operation is that only a single fluid bed is required with consequent savings in capital costs and reduction in operational problems, inasmuch as no sand circulation or replacement is necessary. Further, the heat of reaction for biomass pyrolysis carried only to the point of producing a high proportion of liquid products is small, and may be either slightly endothermic or exothermic depending on the biomass species. As a result, relatively little heat input is required; in general, this can readily be accomplished by heating the fluidized gas to somewhat above the reaction temperature.

In the present work, a cylindrical reactor was used, but there are potential advantages in using other configurations. The residence times in the reactor are difficult to define. In this work, the residence time is calculated as an apparent gas residence time, based on the volumetric flow rate of the total fluidizing gas at reactor conditions and the net empty reactor volume. In fact, actual vapour residence time will be less than this, because of the vapour volume generated in pyrolysis. This extra flow would reduce the apparent vapour residence time by about 15% on the average. Char residence time is not known with any certainty, but the small char accumulation in the bed at any time suggests that it would be only slightly longer than that of the vapour.

Temperatures and pressures were monitored throughout the unit, the former by means of thermocouples, and the latter by differential and absolute bellows-type gauges. The fed hopper and feeder were mounted on a hinged platform, resting on a load cell, and hopper weight loss was recorded continuously. Reaction temperature was controlled by a thermocouple in the fluid bed that regulated the inlet gas heating coils.

Normally, a run was started by bringing the reactor to temperature with total recycle and nitrogen flowing in the system. After reaching a suitable temperature level, biomass feed was started (from 1.5 - 3.0 kg/hr) and within about 10-20 minutes, the system reached a final steady state with respect to both temperatures and gas composition (system volume was small so total system residence time was about 10 seconds). Run time varied from 45 minutes to 120 minutes, and temperatures from 425°-625°C. Reactor pressure was usually about 125 kPa absolute.

No chronic problems were encountered with the pilot unit except for the need to remove all of the tar aerosols, which form readily in the condensers, in order not to affect compressor operation. The amount and nature of the tar aerosol formed depends on the type of cooling used. In this unit, surface condensers were used to allow complete material balances to be carried out. As a result, a considerable amount of stable aerosol formed, the amount of which also depended on the type of biomass.

Analysis of Products

In general, four products were recovered after a run. From the two condensers, a fairly fluid oil was obtained which also contained most of the water produced. The "liquor" fraction varied from about 50% of the total liquid yield for wheat straw to about 85% for the poplar-aspen and maple. It was a dark single-phase fluid with an acrid smell, and was easily pourable. It contained from 15%-30% water, depending on the type of feed and moisture content. This oil appeared to be stable on the shelf over periods of weeks, if not exposed to air.

After a run, the condensers were washed with acetone, the solution filtered, and t acetone subsequently evaporated under vacuum to yield a "tar" fraction. Additional tar was collected in the filters, which were also weighed before and after a run. If the tar in the filters was substantial, which was the case only at the highest temperatures or for wheat straw feed, the filters were also washed with acetone and the tar recovered. All liquid products were readily soluble in acetone.

Char was collected from the char pot and weighed. This char was very reactive and precautions were required to ensure that it did not react with air when hot. The residue from the filtration of the tar fraction (that is, any acetone-insoluble material recovered) was also included in the char fraction.

Gases were analyzed for CO and CO_2 in an on-line, infra-red gas analyzer-recorder. Also, samples of the product gas were taken periodically and analyzed by gas chromatography.

Biomass feed was quantified not only from the weight loss recorder but was checked by independent weighing of the feed hopper contents before and after a run, too. Elemental analyses of feed and products was done using a Perkin-Elmer 240C elemental analyzer. Water content of liquid products was determined by Karl Fischer titration.

RESULTS AND DISCUSSION

The feed used in all cases, unless otherwise specified, was~595 μm (~30 mesh) in size. No fines were removed from the feed, so a rather wide range of particle sizes resulted. Particles from wood were primarily cylindrical, with the largest diameter approaching 595 μm, but the length often four or five times that value. The wheat straw had a significant amount of "dust", which was not noticeable with the wood feed material.

Organic liquid and char yields for the pyrolysis of hybrid poplar-aspen are shown in Figure 1 and gas yields in Figure 2. The two feed materials used were from the same experimental plantation at Brockville, Ontario (Ontario Ministry of Natural Resources, Populus canescens, 8 years old, Clone C-147): the wood sample was a clean, bark-free standard lot received from International Energy Agency, and the "whole tree" sample was obtained from six whole trees, including bark and branches. This latter material contained an estimated 15%-20% bark.

Yields of organic liquids from both the clean wood and the whole tree material show very similar values, with the clean wood giving slightly higher maximum yields (about 69% vs 66% of the moisture free wood). The trends and yields of organic liquids follow closely the same trends that were observed in the bench-scale unit, although the yield of liquids is somewhat higher in the pilot plant unit. One test was also carried out using the clean wood and a particle size of -1190 to +595μm. The yields of organic liquids, gas and char are very similar to those obtained with the ~595 μm feeds in the pilot plant of the -250 +105 μm feed used in the bench scale apparatus.

Figure 3 also shows the yields of CO and CO_2 produced by pyrolysis in the pilot plant, as well as the total gas yield. At temperatures above 550° there is a considerably larger gas yield from the pilot plant than that obtained in bench scale work under an N_2 atmosphere. This higher gas yield at higher temperatures corresponds to the more rapid falling off in liquid yield for the pilot plant runs, as shown in Figure 1. It may be concluded that relatively little additional primary pyrolysis occurs in the fluid bed of the larger reactor, since char yields in pilot plant and bench scale units were similar at higher temperatures. Therefore, the extra gas formed and the reduced liquid yield in the pilot plant at higher temperatures represent secondary decomposition reactions, likely occurring in the freeboard space. However, at the optimal temperature of about 500°C, these secondary reactions appear to be minimized.

The gas produced is a medium-quality gas with a higher heating value of about 14.4 MJ/std.m^3 for the composition obtained at 500°C. This heating value increases at higher pyrolysis temperatures as CH_4 content increases and CO_2 content decreases.

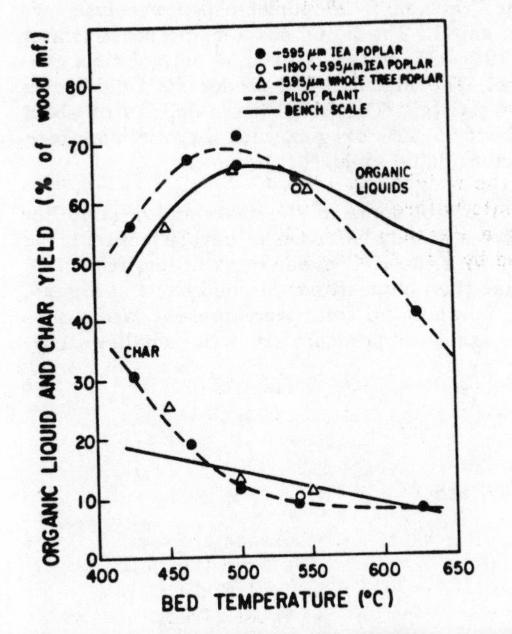

Figure 1: Organic Liquid and Char Yields from IEA and Whole Tree Poplar-Aspen Wood Comparison with Bench Scale

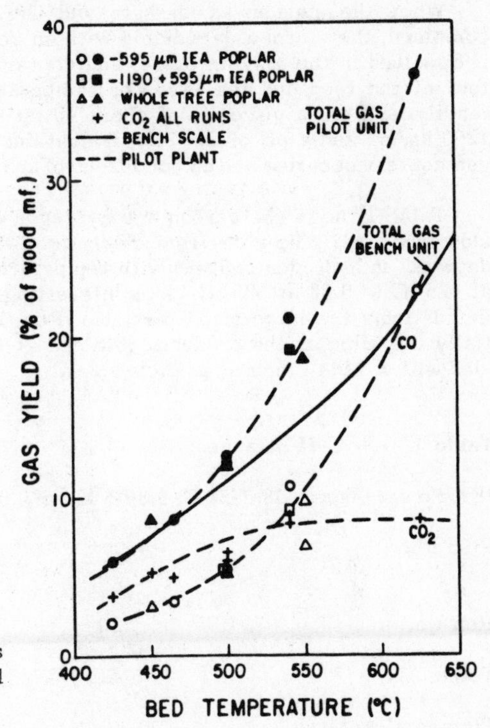

Figure 2: Gas Yields from IEA and Whole Tree Poplar-Aspen Wood Comparison with Bench Scale

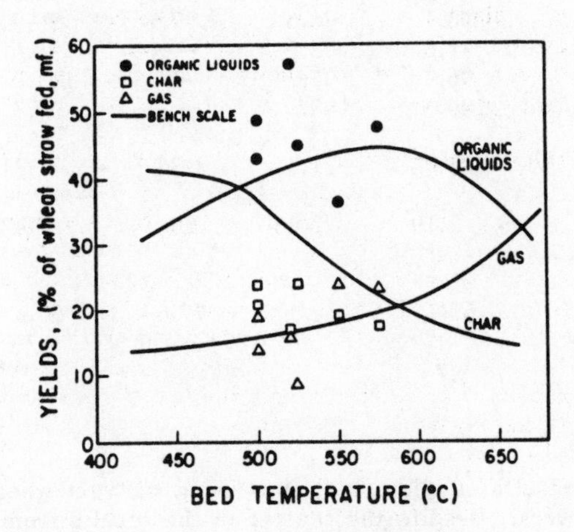

Figure 3: Yields of Liquid, Char and Gas from Wheat Straw Comparison with Bench Scale Results

When the aqueous condensate and the tar fraction from poplar-aspen pyrolysis are combined, they form a dark liquid with an acrid smell. The liquid has a single phase (water is dissolved in the organic phase) and contains 10%-20% water, depending on moisture content of the feed and the pyrolysis temperature. The mixed liquid product is fluid, pours readily, and has a viscosity of about 40 cp at 40°C (4). The liquid has a density of about 1200 kg/m^3 and a pH of 2.3-2.8. It contains about 35-40% oxygen, with the minimum oxygen content occuring at the conditions of maximum liquid yield.

Table 1 shows the carbon and hydrogen contents of some of the chars and organic liquids produced. The char from clean wood shows a steady increase in carbon content and decrease in hydrogen content with temperature, with the H/C molar ratio falling from 1.19 at 425°C to 0.38 at 625°C. It is interesting that the compositions of the char and organic liquids from the large-sized particles (Run 34), in which no fines were present, are essentially the same as the products obtained at the same temperature from the smaller particles with a wide range of particle sizes.

Table 1. C-H Balances

IEA Poplar-Aspen, 48.39% C, 5.89% H (mf), 0.44% ash

-595 μm

Run No.	30	28	27 29	25	26
Temp. °C	425	465	500	541	625
Elemental Analysis %					
Char C	55.13	59.28	66.45	69.99	73.16
Char H	5.48	5.25	4.43	2.80	2.33
Org. Liquid C	53.36	56.61	55.19	57.88	56.59
Org. Liquid H	5.69	6.18	6.1	6.34	6.34
C Recovery %	92.1	98.6	97.5	99.9	86.4
C Normalized %	95.8	98.5	---	98.6	96.5
H Recovery %	96.4	103.2	94.3	101.0	84.1
H Normalized %	100.3	103.1	---	99.7	93.9
H/C molar ratio, liquid	1.28	1.31	1.33	1.31	1.34
H/C molar ratio, char	1.19	1.06	.80	.48	.38
C Conversion to liquids, (norm)	60.1	69.5	77.6	63.2	49.7

* -1190+595μm

Figure 3 shows the results of pilot plant runs using -595μm wheat straw compared to those from bench-scale work. Despite the scatter in the results from the pilot plant tests, the two sets of data are in general agreement. Maximum liquid yields appear to occur at somewhat lower temperatures in the pilot unit, and char yields appear to be less at lower temperatures. However, the considerable amount of very fine dust in the pilot plant feed undoubtedly causes some error in the relative amounts of tar and char reported. The combined oil and tar from the wheat straw was less viscous and had a significantly higher pH (3.2 to 3.8) than did the wood oil. Organic liquid yields were lower and char yields higher from wheat straw than from wood, in agreement with bench-scale results. Tar elemental analyses are very similar for wheat straw and wood. However, the char from wheat straw has a lower carbon content and a much higher ash content than that from wood.

The development of processes based on the thermal pyrolysis of biomass have been directed largely to gasification, either for fuel gas or synthesis gas production, and many such processes have been described. The pyrolysis of biomass to give a maximum direct yield of liquid products has received much less attention, and very few studies have been reported, for example, Kosstrin (1980), Roy et al (1982), Scott and Piskorz (1983). Of these, only the work of Kosstrin, and Scott and Piskorz, appears to have both used a fluid bed reactor and been directed to maximizing liquid yields. Roy et al pyrolysed batch-wise under vacuum, and Duncan et al (hyflex process) used a transport reactor under pressure. Comparison of the results reported here with those of other investigations, whether using fluid-bed reactors or other types, shows that as high or higher organic liquid yields have been achieved in this work. Although there is extensive literature on the mechanisms of biomass pyrolysis reactions, until the heat and mass transfer behaviour in a particular reactor are understood, it is difficult to speculate on why liquid yields may be better in one process than another. In the case of the fluidized bed reactor, this analysis is presently underway in our laboratories.

ECONOMIC ANALYSIS

A preliminary economic analysis has been carried out for a plant processing 1000 t/day (dry basis) of waste wood. The analysis was based on a conceptual flow sheet derived from the process described here; but, it used scrubbers to recover liquid products, rather than condensers. Heat for the reaction was assumed to be supplied by burning the char and a portion of the product gas. Heat was recovered from the scrubbers and char burner, and the waste heat was used for drying of the wood chip feed, assumed to contain 40% moisture. The liquid product was not further treated or separated from the scrubbing liquid in this analysis, as the scrubbing liquid was assumed to be a product fraction.

The total capital cost (fixed plus working capital) was estimated to be $33.8 million (1983), and the total annual production cost, excluding raw material cost, for the process was $55.20/tonne of organic liquid (water-free), based on 330 days/year of operation and a 10-year plant life, without credit for unused fuel gas. Allowing a credit for this gas, assuming a heating value of 19880 MJ/kg for the organic liquid, and excluding raw material cost, an energy cost of 0.2 cents/MJ is obtained ($2.11 per million BTU). To this must be added the cost of the gree, waste wood chips. At about $30 per dry tonne, the production cost would be about double the above figure. The total energy efficiency, defined as useful output energy referred to all energy inputs (raw material plus utilities), is 68% - 70%.

It can be concluded that for a large scale plant, the proposed flash pyrolysis process could produce a liquid fuel that might be competitive with fuel oil at the present time, under the right circumstances. If a much smaller plant is contemplated (e.g. 100 dry tonnes/day), the energy production cost is estimated to be about 30% higher.

ACKNOWLEDGEMENT

The authors are pleased to acknowledge the financial support of the Canadian Forestry Service, Department of the Environment, which made this work possible through the ENFOR program. Peter Majerski's assistance with the design, and operation of the pilot plant is also gratefully recognized.

REFERENCES

Scott, D.S. and Piskorz, J.: 1982a. A low rate entrainment feeder for fine solids. Ind. Eng. Chem. Fund. 21:319-322.

Scott, D.S. and Piskorz, J.: 1982b. The flash pyrolysis of aspen-poplar wood. Can. J. Chem. Eng. 60:666-674.

Scott, D.S., Piskorz, J. and Radlein, D.: 1983. Liquid products from the continuous flash pyrolysis of biomass. Submitted to I.E.C. Prod. Des. & Devlop.

Elliott, Douglas C.: August 1983. Private Communication, Pacific Northwest Laboratory, Richland, Wash.

Duncan, D.A., Bodle, W.W. and Banerjee, D.P.: 1981. Production of liquid fuels from biomass by the hyflex process. Energy from Biomass and Wastes V, Inst. Gas Tech., pp 917-938.

Kosstrin, H.M.: October 1981. Direct formation of pyrolysis oil from Biomass. Proc. Specialists Workshop on Fast Pyrolysis of Biomass, Copper Mountain, Colo. SERI/CP 622-1096, Solar Energy Research Institute, U.S. Dept. of Energy, pp 105-121.

Roy, C., de Caumia, B. and Chornet, E.: February 1982. Liquids from biomass by vacuum pyrolysis -- Production aspects. Proc. Specialists Meeting on Biomass Liquefaction, Saskatoon, Sask. ENFOR Program, Canadian Forestry Service, Environment Canada, pp 57-75.

Scott, D.S. and Piskorz, J.: January 1983. Continuous flash pyrolysis of wood for production of liquid fuels. Proc. Symp. on Energy from Biomass and Wastes VII. Inst. Gas Tech., Chicago, pp 113-1146.

Subject Index

Page numbers refer to the first page of pertinent article.